Energy from Biomass

Solar Energy R&D in the European Community

Series E:

Energy from Biomass

Volume 1

Scientific editing: P. CHARTIER and W. PALZ
Publication arrangements: D. NICOLAY

Solar Energy R&D
in the European Community

Volume 1

Energy from Biomass

Proceedings of the
EC Contractors' Meeting
held in Copenhagen, 23-24 June 1981

edited by

P. CHARTIER
INRA, France

and

W. PALZ
Commission of the European Communities

D. REIDEL PUBLISHING COMPANY

Dordrecht, Holland / Boston, U.S.A. / London, England

for the Commission of the European Communities

Library of Congress Cataloging in Publication Data
Main entry under title:

Energy from biomass

(Solar energy R&D in the European Community ; v. 1)
1. Biomass energy—Congresses. I. Chartier, P.
II. Palz, Wolfgang. III. Series.
TP360.E533 333.95'3 81-14377
ISBN-13: 978-94-009-8562-9 **e-ISBN-13: 978-94-009-8560-5**
DOI:10.1007/978-94-009-8560-5

CIP

Organization of the Contractors' meeting by
Commission of the European Communities
Directorate-General Research, Science and Education, Brussels

Publication arrangements by
Commission of the European Communities
Directorate-General Information Market and Innovation, Luxembourg

EUR 7420

Published by D. Reidel Publishing Company
P.O. Box 17, 3300 AA Dordrecht, Holland

Sold and distributed in the U.S.A. and Canada
by Kluwer Boston Inc.,
190 Old Derby Street, Hingham, MA 02043, U.S.A.

In all other countries, sold and distributed
by Kluwer Academic Publishers Group,
P.O. Box 322, 3300 AH Dordrecht, Holland

D. Reidel Publishing Company is a member of the Kluwer Group

PREFACE

This book is the first volume of a new series on solar energy research and development which is carried out in the European Community. The Commission of the European Community's Directorate General (XII) for Research, Science and Education is currently implementing, on a cost-sharing basis, a solar energy R+D programme through contracts with European industry, research institutions and universities.

This programme covers the following sectors :

Project A - Solar Energy Applications to Dwellings
Project B - Thermo-mech. Solar Power Plants
Project C - Photovoltaic Power Generation
Project D - Photochemical, Photoelectrochemical and Photobiological Processes
Project E - Energy from Biomass
Project F - Solar Radiation Data
Project G - Wind Energy
Project H - Solar Energy in Agriculture and Industry

In the course of the programme the Commission organises each year several coordination meetings to which the contractors within each sector are invited to present details of their work.

As the information presented at these meetings is recent and of high interest for other researchers working in these particular sectors, the Commission has decided to start this new series of publications in cooperation with D. Reidel Publishing Company in order that this new information can be quickly and widely circulated for the benefit of advancement and progress in the whole field of solar energy.

This first volume covers the proceedings of the contractors' meeting on "Energy from Biomass" held in June 1981 in Copenhagen. Approximately half a dozen of this type of proceedings will be published each year from now on; they will mostly concern the areas of "Solar Energy Applications to Dwellings", Photovoltaic Power Generation", Photochemical, Photoelectrochemical and Photobiological Processes" and last but not least "Solar Radiation Data".

I hope that this series will attract the attention it deserves from the solar energy community and can contribute to an early implementation of solar energy.

A.S. STRUB
Head of the Energy R+D Programme
Commission of the European Communities

CONTENTS

INTRODUCTION AND SUMMARY

Introduction

The Commission of the European Communities has currently concluded more than 30 contracts on the research and development of energy from biomass. These contracts run from July 1980 to June 1983. They cover all relevant areas of interest in the field of biomass production and utilisation for energy purposes in Europe.

At the coordination meeting in Copenhagen all contractors presented the status of their work which is reproduced in these proceedings. In the framework of its solar energy R+D programme, there have been other contractors meetings on energy from biomass organised by the Commission since 1977. The previous meetings were held in Brussels, Versailles, Taormina and Amsterdam.

The purpose of the EC coordination meetings is to strengthen cooperation on a European level and across national borders. This is achieved through extensive exchange of information and results, discussions about recent work, new ideas and the tasks that remain.

The meeting was chaired by Mr. Ph. Chartier, Project Leader and Mr. W. Palz, Head of the Solar Energy R+D Programme. Sessions were chaired by the following members of the EC expert group "Energy from Biomass" :

- Agricultural wastes, energy crops, Mr. Hansen
- Forestry, wood wastes, Mr. Morandini
- Algae, Mr. Wagener
- Biomass conversion (biological routes), Mr. Lettinga, Mr. Naveau
- Biomass conversion (thermochemical routes), Mr. King, Mr. Hurand.

SUMMARY ON AGRICULTURAL WASTES, ENERGY CROPS

(by G.K. HANSEN)

Four projects have now commenced: three of them concern growing, harvesting and utilization of energy crops, and one the social-economic and agricultural constraints involved in energy production from straw and animal wastes.

In the catch crop project, a theoretical simulation approach is applied together with field experiments. Some interesting results have been obtained. In experimentation with catch crops it is important to note that the influence of these crops on the main crops has been studied. Established experimental methods of intercropping can be applied and apparently there seem to be many possibilities for growing catch crops and increasing the utilisation of arable land in Europe.

There are many problems in the utilisation of native plant species as their behaviour under cultivation is not known. Hence many more experiments are needed in order to have a better understanding of how to utilise them effectively.

Some economic studies of using straw for cereal drying have now been performed. The costs are compared with the cost of oil and propane as fuel sources and give favourable results.

Long term influence on humus has still to be studied.

There are many problems left which the present EC project, although promising, cannot definitely solve: much more work is required and should be supported.

SUMMARY ON FORESTRY, WOOD WASTES

(by Mr. MORANDINI)

The progress of several projects seems quite encouraging.

Several biomass plantation projects are being implemented so that knowledge and direct experience can be gained.

Biomass yield tables have been established for several species in plantations of fast growing species and of coppices.

Progress in the field of development of special harvesting equipment is not well advanced, especially for wood wastes collection in the forest. It has been recommended to better coordinate the work on short rotation forestry which is in progress in the different EC member states.

More attention should be paid to financial and energy balance of short rotation forestry for energy purposes as well as the associated establishment, management, harvesting problems.

SUMMARY ON ALGAE

(by K. WAGENER)

The most challenging aspects of algae for energy farming are clearly their outstanding productivity and their potential to be grown outside arable land. The five activities actually run under EC contracts concern three types of growth system: tubular systems, ponds, open sea mariculture.

The tubular system offers the most perfect control, even sterile conditions can be maintained. Optimal parameters (except dependency on sunlight) guarantee maximal productivities. This system is ideal for biological screening and fundamental physiological studies. For large-scale energy farming the capital investment is too high so that for a profitable basis high-priced valuable products have to be the target. The actual purpose is to study and optimize cultural conditions for Botryococcus braunii and Porphyridium cruentum. These strains are producing extra-cellular biomass, i.e. hydrocarbons or polysaccharides respectively. The aim is to grow them in large-scale systems and on a cost-effective basis.

Ponds are used for mariculture on land. Outdoor ponds for testing various types of algae have been continuously in operation for 2½ years now in Lamezia/Calabria/Italy. The average productivity over the whole time is 60-65 t dry matter/ha/yr; for the more favourable summer months (May - September) it is 90 t/ha/yr. This experience shows that high yields can be obtained over long periods for microalgae. Harvesting by settling or straining has been tested successfully.

Open sea mariculture is presently under investigation in the UK, notably in a low-technology design. This has the advantage of being substantially less capital intensive than former systems by which seaweed is grown in artificially upwelling water. To date, drag force studies on the supporting rope system and preliminary productivity investigations have been carried out.

SUMMARY ON BIOLOGICAL CONVERSION ROUTES

(by Mr. HURAND, Mr. KING)

The presentations and the discussion showed once again the quality of the contractors in the thermochemical conversion programme; they are innovative but also bring hard-headed realism to the problems.

For the production of heat, mainly for use at farm level, straw is an important resource in Europe and is already widely used for energy production. However, one of the main problems relating to small burners is their relatively low efficiency. A Danish project concerns heat recovery in a scrubber, and shows that very good performance is feasible; this kind of equipment will certainly have great developments in the future.

Another way to improve boiler efficiency, storage and transportation costs, was proposed with an interesting concept of hot granulation to prepare the product. This process is said to be less energy-consuming.

The range of thermochemical processes for gasifying biomass include studies on pyrolysis and steam and oxygen gasification using various configurations; heating modes include partial combustion in-situ, re-cycle of combusted product gases.

Although most projects are developing processes which will need to operate at a scale of several hundred tons/day of product, one, at least (Imbert-Werner) is investigating economic ways of up-grading wood in the forest, through production of char and pyrolytic oils. It is envisaged that these materials would be taken to a central plant for further processing, but it will be important to establish the trade-off between the higher cost of processing and the savings in transport costs.

Overall, much progress has been made since the last contractors meeting; the projects are beginning to elaborate laboratory results, and some test-rigs are already operating, whilst others are likely to be commissioned in the next couple of months.

Most projects, however, are still in their early stages so that even basic experimental information is not yet available on the balances for mass, energy and costs.

SUMMARY ON THERMOCHEMICAL CONVERSION ROUTES

(by Mr. LETTINGA)

Whether ethanol production as energy carrier might become attractive for EC-countries is still quite questionable, but the results of research here might become important for countries where presently the production of alcohol as fuel looks attractive.

An important point in enzymatic hydrolysis is to apply the enzymes at an optimal pH, which seems to be in acid pH-region. More work has to be done in this field, also with respect to antagonistic effects. Their findings may become of importance for application of two step bio-conversion routes, i.e. with a separate liquefaction step, either for CH_4- production or for ethanol production.

Presently the use of enzymes is not economical. However the situation may change in the future once it can be shown that the enzymes supplied to the system can be recovered.

Future work should be devoted to this aspect.

Concerning anaerobic digestion, there is contract work on one step digestion processes and two step digestion processes, with a separate liquefaction step.

In the digestion of complex solids wastes a two step approach with a separate liquefaction step, might become attractive, especially if the liquefaction could be improved by supplying specific enzymes.

On the other hand, it still has to be proved whether or not a two step process is more effective and more economical than the one step approach.

Concerning the application of thermophilic digestion, this process might become especially attractive in the treatment of huge quantities of liquid agricultural industrial wastes, presently available at the proper temperature.

SUBJECT I. AGRICULTURAL WASTES, ENERGY CROPS

Energy production using straw and animal wastes as feedstocks - Analysis of both the agricultural (humus) and socio-economic constraints involved

An experimental assessment of native and naturalised species of plants as renewable sources of energy in Great Britain

Studies on the yield of catch crops grown for fuel

Joint research on Arundo Donax as an energy crop

ENERGY PRODUCTION USING STRAW AND ANIMAL WASTES AS FEEDSTOCKS

ANALYSIS OF BOTH THE AGRICULTURAL (HUMUS) AND SOCIO-ECONOMIC CONSTRAINTS INVOLVED

Contract number : ESE-R-014-F

Duration : from July 80 to June 83

Total budget : 1 750 000 FF. CEC contribution 40 %

Head of project : Dr. P. CHASSIN

Contractor : INSTITUT NATIONAL DE LA RECHERCHE AGRONOMIQUE (INRA)

Address : PARIS 149, rue de Grenelle FRANCE

Summary

The use of crop residues for energetic purposes may increase the unstability of the farm production systems by changing the production capacity of soil (soil fertility) and the socio-economic equilibria (financal equilibrium,maintenance of a rural population chiefly in marginalised area...).

Different French INRA (Institut National de la Recherche Agronomique) teams (Avignon, Bordeaux, Chalons sur Marne, Dijon, Laon, Versailles, Grignon and Montpellier) essentially explored two directions :

1) Agronomic aspect : the studies done within the "soil protection under maximum removal of organic matter" programme were aimed at answering the two following questions :

1. - Can the fertility of soils bearing a succession of essentially cereal crops be maintained if nearly all their aerial parts are systematically removed ?

2. - If not, what are the remedial techniques or the minimum amounts of residues which should not be removed so that the negative effects of such a technique be compensated ?

2) Economic aspect : we must know the effect of energetic use of straw at the farm or at the surroundings of the farm on the economic equilibrium of the farm production systems. So we have to perfect the proper methods of economic valuation.

1. Introduction

In the very large field of alternative energy research, many people considered using crop residues for energy rather than for agronomic purposes (soil organic matter control). However, this project may be realistic only if the energy source is renewable, i.e. if the removal of crop residues does not produce a strong decrease in soil fertility, which has not been proved as of today.

Besides appropriated methods mus be elaborated for choosing from different economic points of view the best suited energetic paths wich will be developped at the farm or at the surroundings of the farm.

2. The principle results :

2.1 Agronomic research :

2.1.1 Results of long range experiments

Results from experiments conducted in France suggest that straw incorporation in soil leads to an easier use of soil and a smaller risk of accidents rather than a significant increase in the optimum production capacity of soils (See report n° 3 "Soil protection under maximum removal" contract 05376 ESF).

2.1.2 Estimated decay of organic matter contents.

Traditional concepts are contradicted by the following evidence : existence of a marked biodegradability of maïze stalks (REMY, MULLER) on the one hand, and unlikehood of a "non decomposable" fraction of bound organic matter being proportional to the clay content (MONNIER and STENGEL) on the other and.

The considerable effect of temperature on the in-vitro straw decaying rate (29 % at 10°C against 70 % at 25°C in a year) is likely to account for the relative lack of interest for high straw exportation shown by agronomes in nothern countries. It is also to be noted that nitrogen supply decreases the carbon mineralization rate (REMY-MULLER).

2.1.3 Laboratory studies

MONNIER and GUERIF have shown that properties such as elasticity or sensitivity to soil compaction are strongly dependent on the presence of straw fragments. Straw incorporation in soil thus improves the bearing capacity and trafficability of soils.

Besides, JUSTE and TAUZIN have shown that straw decay in soils which are poor in one trace-elements (zinc) due to agricultural techniques, tends to compensate for this defect.

2.1.4 Study of the fertilizing value of ashes

The above two examples clearly show the necessity of subordinating the study of remedial techniques to the indeniable existence of straw action on soil properties. Bearing this in mind, JUSTE and TAUZIN yielded evidence that the supply of fertilizing elements by straw may be compensated for by restituting ashes produced during straw combustion. They also show in an indisputable way that the potassic fertilizing value of ashes is slightly or much higher than that of K_2SO_4.

2.2 Economic aspect

2.2.1 The advantage of straw

Straw is interesting for heat production by direct burning according to different paths :

- at the farm : for domestic heating, grain drying, air-conditionned breeding buildings, glass-houses heating.

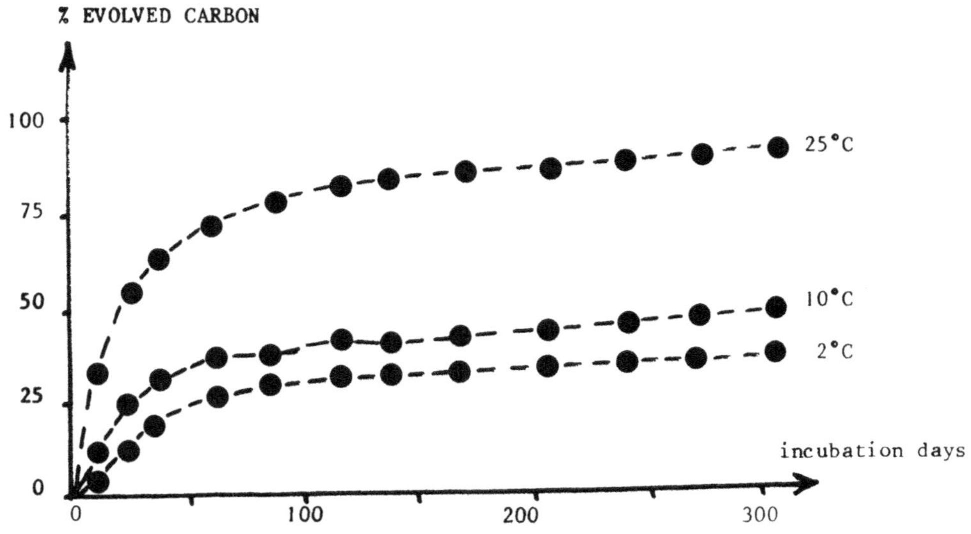

KINETIC OF CORN-STALK MINERALIZATION IN LAOMY SOIL

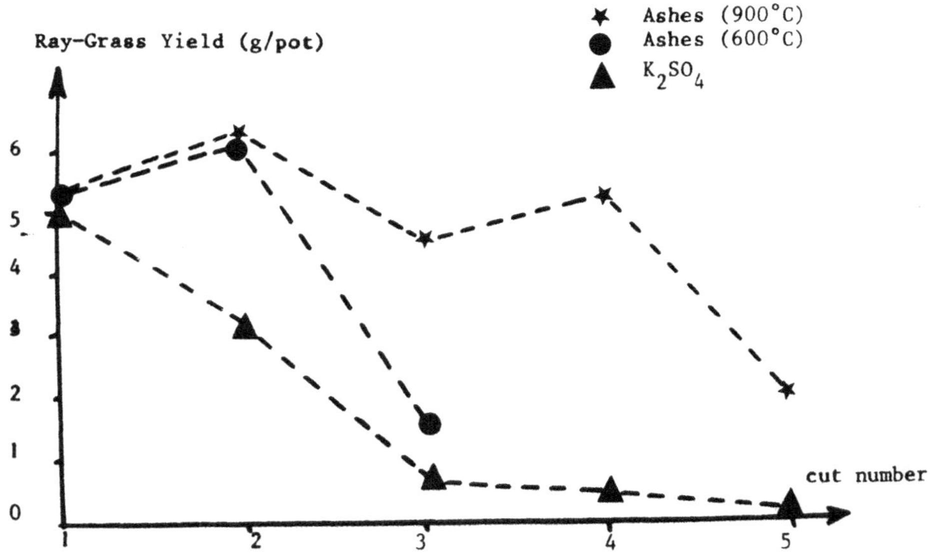

FERTILISING VALUES OF ASHES PRODUCED IN STRAW FURNACES

-At the farm surroundings : for drying different farm products (lucern, pulps...) and as, straw pellets reused by peasants or rural couples.

As an example of our research we showed that in subhumid region and with a "corn-wheat-wheat" system and if the cost of fuel is increasing

- by traditional drying methods, corn disappears
- by drying with straw, corn is maintaining.

In that case the drying operation needs 15 to 20 kg of straw for one quintal of dried corn, wich means that about 15 to 20 % of straw production is nessary to dry the corn produced (output of 55q/ha of corn and of 3 tons of straw per hectare.

In a first stage, about 5 millions of tons from 25 millions produced, wich constitutes the most economic part of the resource, ought to be converted, without an accute competition with traditional agronomic and zootechnic uses. Agronomic research will allow to precise this scale of sizes.

2.2.2 Simultaneously methodologic research were conducted :

They allowed us to determine and to criticize some criterions of economic analysis of individual or collective projects using straw as an energetic material.

- the payback period (payout period) test.
- the discounted cost test.
- the discounted rate of per hour family labour income, which means the introduction of social elements in the analysis.

For the two last criterions impacts of the use of straw for energy requirements on the technical equilibrium of the farm production system are taken intoaccount : for instance :

. the correction of the disturbed mineral balance of soil with fertilizers,
. the cost of the palliatives to correct the organic balance of the soil,
. the cost of temporary labour, the cost due to the backwardness of some agrilcutural operations...

3. Conclusion

Concerning the agronomic part it can be noticed that the research programme includes three aspects :

- estimation of quantitative, qualitative and spatial evolution of organic matter as a function of the crop residue control.
- further studies concerning the action of organic matter on soil properties and behaviour.
- testing of remedial techniques.

Thus, we may think that this comprehensive knowledge will allow us to determine a certain number of rules to comply with so that the use of crop residues for energy purposes does not cause a strong decrease in the soil agricultural production capacity.

In regards to the economic part we have to perfect two points :

- the use of system analysis for a good estimation of the insertion of different energy paths best suited to the rural constraints, chiefly for the marginalized area where the main problem is to maintain a certain level of population : (each new technology may make disappear the unstable farms...) and where some ecologic impacts (unstability of soils...) must be taken in account.
- the estimation of regional competition between the different paths of ernergetic valorization of straw and animal wastes to the amount of 5 millions tons has also to be perfected, i.e.
  . competition between the different energetic paths,
  . competition of the energetic paths with traditional uses (agronomic, zootechnic),
  . competition of the energetic paths with industrial uses (fibres, pulp).

Influence of an increasing cost of energy on the course rotation of a cereal system, according to the drying method chosen for corn (drying operation at the farm).

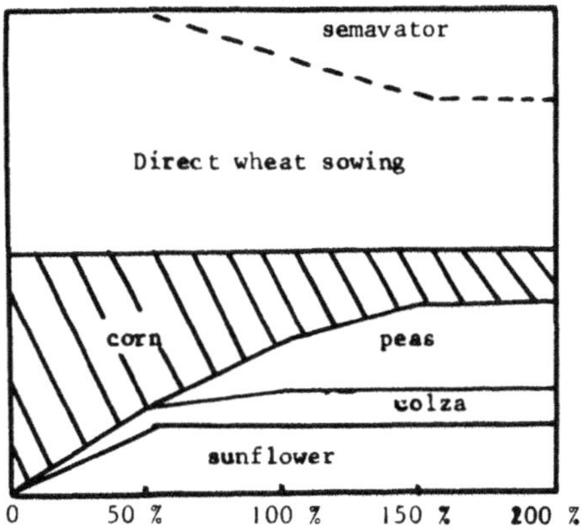

DRIER WITH FUEL

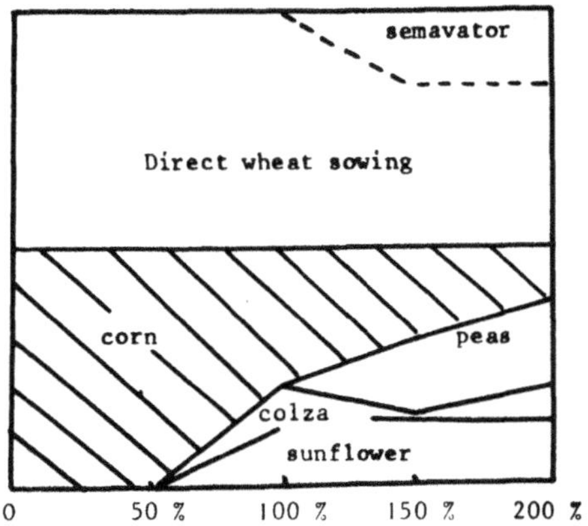

DRIER WITH FUEL-PROPANE

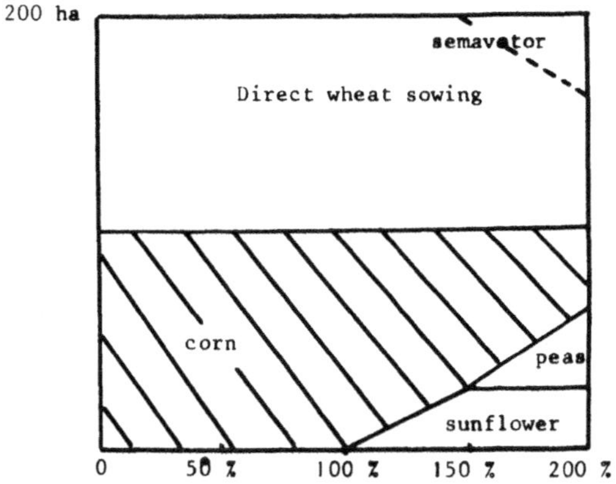

DRIER WITH BOTTLED STRAW

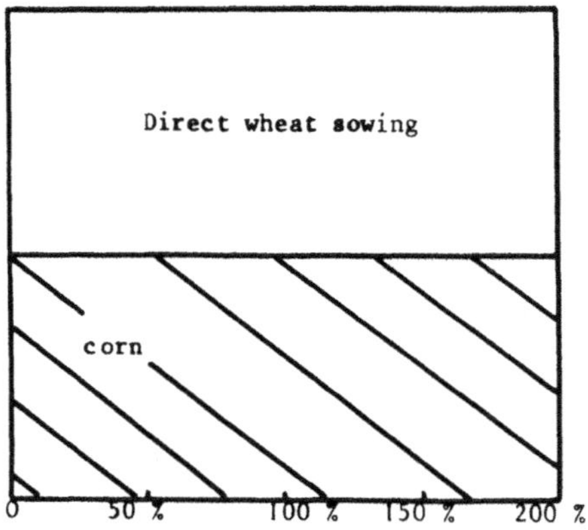

DRIER WITH STRAW IN BIG BALES

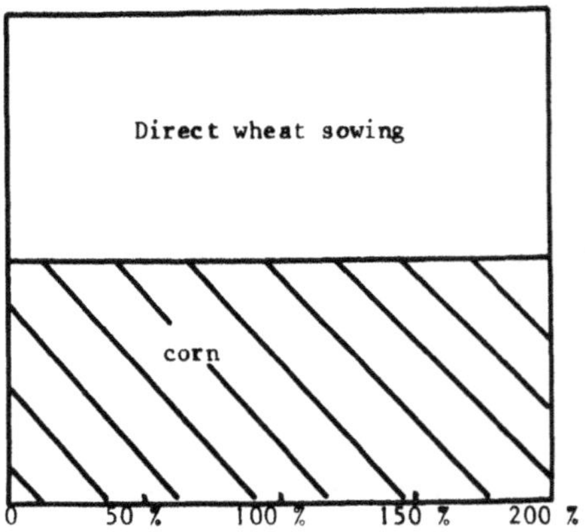

DRIER WITH CHOPED STRAW

"AN EXPERIMENTAL ASSESSMENT OF NATIVE AND NATURALISED SPECIES OF PLANTS AS RENEWABLE SOURCES OF ENERGY IN GREAT BRITAIN"

Authors :	T.V. CALLAGHAN, R. SCOTT and G.J. LAWSON
Contract Number:	ESE-R-015-UK
Duration:	36 months 1 July 1980 - 30 June 1983
Total Budget:	£254,826 CEC Contribution : £48,000
Head of Project:	Dr. T. V. Callaghan, Institute of Terrestrial Ecology
Contractor:	Natural Environment Research Council's Institute of Terrestrial Ecology
Address:	Natural Environment Research Council, Polaris House, SWINDON, Wilts. SN2 1EU.

Summary

Some wild plants of Great Britain could be used for energy sources because they cover vast areas or are very productive. However, no information is available on the stability of yield under various harvesting strategies or weather conditions, and no attempt has been made to increase the yields of these "weeds" by adding fertilizers.

This report presents results from the first year of a three year experiment to study the effects of repeated harvesting, the date of harvesting and the addition of fertilizers on the yields of *Pteridium aquilinum, Spartina anglica* and *Reynoutria japonica*. It also describes parallel experiments on the independent effects of nitrogen, phosphorus and potassium, and 4 planting densities on the yields of these and other potential energy crops.

P. aquilinum, which covers 3,200 km^2 of Great Britain, could be harvested in summer to give *ca.* 9 t ha^{-1} which would be digested anaerobically to give 27 m^3 of biogas t^{-1} day^{-1}. *P. aquilinum* harvested in Autumn, and *S. anglica* harvested in winter (with yields of *ca.* 14 t ha^{-1}) would enter a thermal conversion route. Yields of a recently established monoculture of *R. japonica* were low (0.8 t ha^{-1}) due to a low planting density and abnormal weather conditions but survival was good and yields should improve. No early season responses to fertilizers were shown but re-growth after early harvests was correlated with fertilizer input.

Results from the next two years will be critical in relating the stability of yield to harvesting pressure and identifying ways of increasing yields.

1. INTRODUCTION

Some native and naturalised plant species of Great Britain show considerable potential as energy crops (1,2). Measurements of yields and estimates of distribution and extent (3) show that some species currently occurring over wide areas could be harvested immediately with little disturbance to traditional land use whereas scarce but productive species could be planted on good agricultural land. Widespread species would be managed as opportunity energy crops and currently scarce species as dedicated energy crops. Of the species studied (3) *Pteridium aquilinum* and *Calluna vulgaris* have the greatest potential as opportunity energy crops whereas *Reynoutria japonica, R. sachalinensis* and *Impatiens glandulifera* could be dedicated energy crops. *Spartina anglica* also showed promise as an energy crop.

However, no information is available on the effects of repeated harvesting on subsequent yields of these species or the possibility of increasing their current yields by applying fertilizers. Indeed, attempts have usually been made to eradicate or control these species because of their prolific growth. Also, little is known about their yields in relation to weather conditions, different habitats, geographical locations and genetic differentiation.

To answer these and related questions, a series of three-year experiments has been initiated. This report presents experimental results obtained at the end of the first year of the trials, together with a description of other recently established experiments.

2. EXPERIMENTAL APPROACH

2.1 Major Factorial Experiment

A standardised factorial experiment on *Pteridium, Spartina* and *R. japonica* was designed to test the effects on yield of:

a) applications of a standard 20:10:10 mixture of N, P and K at 0, 0.5, 1 and 2 t ha^{-1} (FO, 1, 2 and 3 in Figs. 1 and 2)

b) harvesting on one of four occasions in the season

c) harvesting once, twice or three times over a three year period.

The experiment also allows for the measurement of photosynthesis, respiration, translocation, chemical composition of shoots and within-season re-growth after harvesting (4).

At each harvesting date the vegetation is cut to ground level, collected, sorted and oven dried (80° C to constant weight). A sub-sample of the weighed material is analysed for chemical composition while some fresh material is sent to University College, Cardiff for anaerobic digestion trials. At the end of each season re-growth is estimated by counting stem density and weighing individual stems from a buffer zone surrounding samples squares (4). Crop development, i.e. height, density and leaf area is also assessed by sub-sampling.

2.2 Independent Nutrient Trial

Experiments were initiated in Spring 1980 to separate the effects of N, P and K on the yields of *Pteridium, Spartina* and *R. japonica* by varying applications of N, P and K independently. The three levels of P_2O_5 and K_2O were 0, 15 and 30 g m^{-2} whereas N was applied at 0, 30 and 60 g m^{-2}. This experiment will be harvested once, during Autumn 1982.

2.3 Plant Density Trial

No data exist on the establishment of monocultures of natural energy crops. The yield of these crops will depend upon the speed with which a

TABLE I

A summary of the potential of native and naturalised plant species as energy crops

THE RESOURCE	RESOURCE CHARACTERISTICS	USE OF RESOURCE	SIZE OF RESOURCE	RESEARCH REQUIREMENTS
NATIVE AND NATURALISED PLANT SPECIES	WIDESPREAD AND PRODUCTIVE (*Pteridium aquilinum, Spartina anglica*) extent: 120; 3,200 km^2 yield: 8.7-15.8 t ha^{-1}	OPPORTUNITY ENERGY CROPS land use relatively unaltered: or encouraged spread: or planned control. Conversion: anaerobic digestion or thermal.	3,200 km^2 at 8.7 t ha^{-1} (fresh feedstock) = 2,784,000 t or at 7.7 t ha^{-1} (dead feedstock) = 2,464,000 t	Response to harvesting pressures; fertilizers, variation in weather; suitability for conversion.
	VERY WIDESPREAD BUT NOT PRODUCTIVE (*Calluna vulgaris*) extent: 14,900 km^2 yield: 1 - 2 t ha^{-1}	OPPORTUNITY ENERGY CROPS land use unaltered Conversion: thermal	14,900 km^2 at 1-2 t ha^{-1} = 1,490,000 - 2,980,000 t of dead feedstock	
	SCARCE BUT PRODUCTIVE (*Reynoutria japonica, R. sachalinensis, Impatiens glandulifera*) extent: < 10 km^2 yield: 10-25 t ha^{-1}	DEDICATED ENERGY CROPS Plants collected, propagated and new crops established. Conversion: mainly thermal methods.	Assume railway embankments and road verges available initially: 250 km^2 at 10 t ha^{-1} (dead feedstock) = 250,000 t	Methods of propagation density of planting, period of establishment, response to harvesting pressures, fertilizers, variation in weather and location; suitability for conversion; susceptibility to disease and pests.
	RELATIVELY UNKNOWN SPECIES (*Gunnera manicata*) extent: ? yield: ?	POTENTIAL DEDICATED ENERGY CROPS Plants collected, propagated and new crops established. Conversion: anaerobic digestion or thermal methods.	Size of resource unknown.	

Note: all weights are tonnes dry matter

leaf canopy closes which is related to plant density. A trial was therefore, initiated in Spring 1981 to:

a) investigate the effects of plant density on the yields of *Reynoutria japonica* and *R. sachalinensis*.

b) compare the yields of *R. japonica* and *R. sachalinensis* over a range of plant densities.

Both species were planted equidistantly at 4, 9, 16 and 25 plants m^{-2}. The experiment will be assessed during Autumn 1982.

2.4 The Collection and Establishment of Fast Growing Species

Little is known about the yields of faster growing invasive species or common and exotic garden plants (1). Because these species represent an unknown pool of potential energy crops, a collection is being established of species which appear to be productive. This collection will provide material for propagation and includes *Gunnera manicata, Impatiens glandulifera, Reynoutria sachalinensis, Smyrnium olusatrum, Galega officinalis* and *Lupinus arboreus*.

3. RESULTS

3.1 *Pteridium aquilinum*

The seasonal development of *Pteridium* was not significantly (P>.05) affected by the four fertilizer treatments as the above ground growth of perennial species may depend upon reserves of nutrients and carbohydrates stored in rhizomes. Yields in future years should respond to initial fertilizer treatments.

The dry weights of *Pteridium* fronds reach 8.7 t ha^{-1} by the beginning of September (Fig. 1a) but yields of 7.7 t ha^{-1} could be obtained from senescent fronds later in the growing season. A water content of 67% of the fresh weight during September indicates that 26 t ha^{-1} of fresh material would be harvested and transported if fronds were cut at this time for processing in an anaerobic digestor.

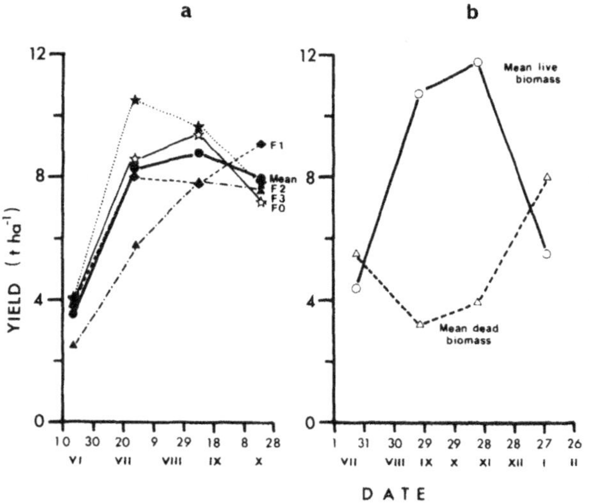

Fig. 1. Yields (t ha^{-1}) of (a) *P. aquilinum*, (b) *S. anglica* and (c) *R. japonica* during 1980. (F0 to F3 = fertilizer treatments)

The amount of re-growth was obviously greater in quadrats which had been cut earlier in the season but the density of fronds emerging after a harvest in July was strongly correlated with nutrient treatment ($P \leq .001$; Fig. 2a). Stem density after a late cut, and with the highest fertilizer application, was similar to that following an earlier harvest and longer growing season (Fig. 2a). This, again, suggests a delay in the effect of fertilizers on this perennial species.

3.2 *Spartina anglica*

Like *Pteridium*, *Spartina* develops late in the season but *Spartina* continues to grow during the winter. A maximum leaf area index of 10.7 m^2 m^{-2} is achieved in early September.

Yields of *Spartina* assessed in one year may include dead leaves from previous years. However, a maximum dry weight of green tissues of 11 t ha^{-1} was obtained in November when dead tissues weighed 4.8 t ha^{-1} (Fig. 1b). Unfortunately, the 4.8 t ha^{-1} cannot be apportioned between old and new leaves until next year, but initial yields of between 11 and 15.8 t ha^{-1} of predominantly green material may be assumed for *Spartina* harvested in November. Yields up to 13.7 t ha^{-1} of predominantly dead material could be obtained in January. High water contents of up to 80% fresh weight reflect fresh weights of 67 t ha^{-1}.

The effects of fertilizer treatments on shoots produced at the beginning of the growing season were insignificant ($P > .05$) but, after an early cut, fertilizer treatments begin to effect the growth of later shoot generations (Fig 2b).

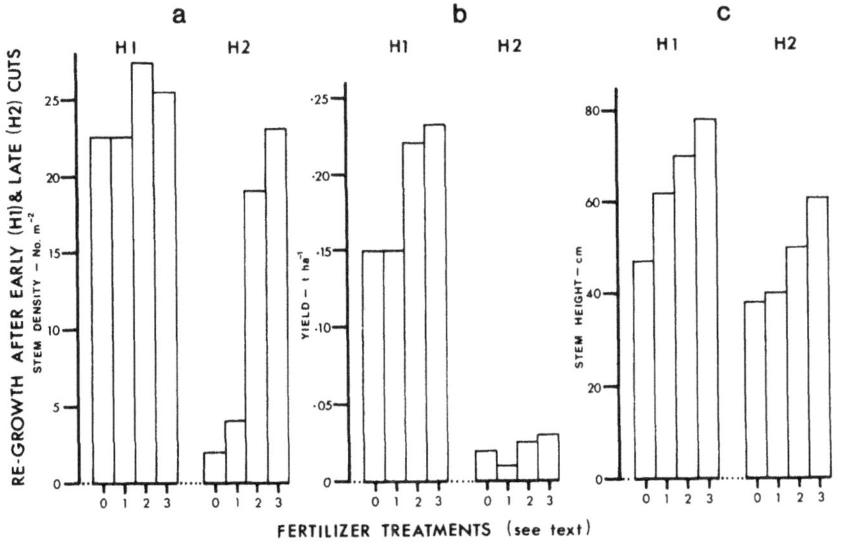

Fig. 2. The effect of fertilizer treatment on re- growth of (a) *P. aquilinum*, (b) *S. anglica* and (c) *R. japonica*. (see text)

3.3 *Reynoutria japonica*

Rhizome fragments of *R. japonica* (mean weight: 62.6 ± 8.8 g; mean number of buds: 4.03 ± 0.45) were planted by hand(4 m^{-2}). Survival was good as stem density reached 7.5 m^{-2} within 5 months of planting. The subsequent development of plants was less than in well established stands (3). Stem height reach only 67 cm compared with 188 cm at a native site.

Similarly, yields were much lower at the transplant site (0.8 t ha^{-1}) than at native sites (9.8 to 11.4 t ha^{-1}) (Fig. 1c).

There are three main causes of low yields at the transplant site: shoots were harvested before the underground system had developed, the density of planting was low and a drought was experienced within two months of planting un-rooted rhizome fragments. Poor development of *R. japonica* allowed the growth of other species which increased site yields to 6.0 t ha^{-1}. However, *ca.* 100% survival of *R. japonica* under adverse weather conditions gives rise to optimism for higher yields in future, particularly if plant density is increased. Increased re-growth on fertilized areas also suggests optimism for future yields (Fig. 2c).

4. DISCUSSION

The results of the first year of the three year experiments provide only limited data on the management and long term stability of potential energy crops but they provide valuable confirmation of results from earlier field trials.

In the present experiments, *Pteridium* yielded up to 8.7 t ha^{-1} while the maximum yields at two other sites were 8.7 and 8.9 t ha^{-1}(3). It would appear that *Pteridium* can yield 8.7 t ha^{-1} or more over a range of conditions and 7.7 t ha^{-1} when senescent.

Preliminary yields of biogas after anaerobic digestion gave 27m^3 of biogas per tonne of feedstock but the burning properties of *Pteridium* remain to be determined. If a yield of 8 t ha^{-1} is assumed, the cost of energy from *Pteridium* would be *ca.* £1.60 per barrel oil equivalent (Table II). *Pteridium* could, therefore, help the poor economy of the uplands by using off-peak labour and equipment and ending the current costs of control measures.

TABLE II

The cost of energy at the farm gate from *Pteridium*

Standing crop of bracken	= 8 t ha^{-1}
Energy content of bracken	= 144 GJ ha^{-1}
Cutting rate (including breakages)	= 3.98 ha day^{-1}
Cutting and transport cost per unit time	= £152 day^{-1}
Cutting and transport cost per unit area	= £152/3.98 = £38.19 ha^{-1}
Cutting and transport cost per unit energy	= £38.19/144 GJ = £0.27 GJ^{-1}
	= £7.27/tonne coal equivalent
	= £1.60/barrel oil equivalent

The first year's results on *Spartina* were also optimistic. Predominantly living material yielded 15.8 t ha^{-1}, while winter biomass yielded 13.7 t ha^{-1}; earlier work(3) recorded yields of 16.8 and 11.4 t ha^{-1} for living and dead material respectively. Thus, *Spartina* is a productive species growing on land for which there are few alternative uses; its introduction into other similar areas should be considered.

Initial yields of *R. japonica* were disappointing but survival of plants was surprisingly good. This survival and successful re-growth after harvesting suggests that yields will increase markedly once rhizomes have become fully established, particularly if they are planted at greater densities.

Understandably, no attempts have previously been made to improve the yields of the weed species studied here, but an increase in yield after fertilizer applications was expected. None of the perennial species showed immediate increases of yield after fertilizers had been applied.

Shoots produced in Spring appear to depend on nutrients and energy stored in rhizomes formed in previous years but shoots produced later in the season respond to fertilizer treatments applied in Spring. The next phase of the experiment will show if fertilizers have a significant longer term effect on yield.

5. CONCLUSION

It is difficult to draw major conclusions from the first year's results of a three year experiment. However, initial yields of *Pteridium* and *Spartina* suggest optimism for their use as energy crops while the ease of establishing monocultures of *R. japonica* has been demonstrated. Future results will show if the high yields of *Pteridium* and *Spartina* can be maintained under various harvesting regimes - or even improved by adding fertilizers - while providing yields of the developing monocultures of *R. japonica*. Parallel experiments which have been successfully established will indicate optimal densities of planting for *R. japonica* and *R. sachalinensis*, and will provide data on the yields of small monocultures of *R. sachalinensis* and *Impatiens*, and the interactions of N, P and K with the yields of these species.

The three species studied most intensively would not be mutually exclusive energy crops. The upland and heathland habitats of *Pteridium* in Great Britain (3,200 km^2), the coastal habitat of *Spartina* (120 km^2 currently colonised) and the railway embankment/roadside verge habitats which could be planted with the *Reynoutria* species (250 km^2) are not interchangeable but together could contribute a considerable renewable feedstock resource for conversion to biofuels. If an average yield of only 5 t ha^{-1} could be sustained over this large underutilised area 1,785,000 t yr^{-1} of feedstock would be available without significantly affecting current land use. This feedstock would contain 32 PJ or 1.2 million tonnes of coal equivalent and excludes a possible 27-54 PJ from *Calluna*. This large unutilised resource should not be neglected.

6. REFERENCES

1. CALLAGHAN, T.V., MILLAR, A., POWELL, D. & LAWSON, G.J. (1978). Carbon as a renewable energy resource in the UK - conceptual approach. In: Fuels from Biological Materials Programme. Report to UK Dept. of Energy, Project 2, Cambridge. Institute of Terrestrial Ecology. 165 pp.
2. LAWSON, G.J., CALLAGHAN, T.V. & SCOTT, R. (1980). Natural vegetation as a renewable energy resource in the UK. In: Fuels from Biological Materials Programme. Report to UK Dept. of Energy, Project 7, Cambridge. Institute of Terrestrial Ecology, 177 pp.
3. CALLAGHAN, T.V., SCOTT, R. & WHITTAKER H. A. (1981). The yield, development and chemical composition of some fast-growing indigenous and naturalised British plant species in relation to management as energy crops. In: Fuels from Biological Materials Programme. Report to UK Dept. of Energy. Cambridge. Institute of Terrestrial Ecology. 178 pp.
4. CALLAGHAN, T.V., SCOTT, R., LAWSON, G.J. & WHITTAKER H.A. (in press). An experimental assessment of native and naturalised species of plants as renewable sources of energy in Great Britain. In: Project E - Energy from biomass. Proceedings of the coordination meeting of contractors, Amsterdam 1980. Brussels, Commission of the European Communities.

STUDIES ON THE YIELD OF CATCH CROPS GROWN FOR FUEL

Author : S.P. CARRUTHERS

Contract Number : ESE-R-034-UK(H)

Duration : 36 months 1 July 1980 - 30 June 1983

Total Budget : £111,301 CEC Contribution : £48,000

Head of Project : Professor C.R.W. Spedding, University of Reading.

Contractor : University of Reading.

Address : Department of Agriculture and Horticulture,
University of Reading,
Reading, Berkshire,
RG6 2AT, U.K.

Summary

Growing catch crops for fuel is one of a number of biofuel strategies within U.K. agriculture; the main advantage is that the system does not displace food production. Productivity is determined by land-time availability, crop yield and conversion efficiency. Crop yield has been investigated theoretically and experimentally; a descriptive model of the catch fuel crop system has been constructed and the effect of species, seed rate and sowing date on growth and yield investigated in a field experiment. Preliminary analyses reveal a strong effect of sowing date and seed rate at later sowing dates on yield. It is hoped to use data from this experiment and an experiment to be carried out later this year to construct a model to predict catch crop yield in various environments.

1. INTRODUCTION

The growing of catch crops for fuel is one of a number of possible strategies for obtaining biofuels from UK agriculture; the main advantage of the system is that it can be introduced into current cropping regimes without displacing food production. Catch crops are grown in the time period available between the harvest and sowing of main crops and provide a feedstock suitable for anaerobic digestion to produce methane.

The productivity of the catch fuel crop system, defined as the energy output as an available fuel, is determined by land-time availability, crop yield and the efficiency of the conversion process. The objective of this project is to study the potential productivity of catch crops grown for fuel in the UK by a consideration of these three determinants; this paper reports, specifically, on recent investigations of crop yield.

2. METHODS

Catch crop yield is being investigated by a combination of theoretical and experimental studies, the relationship between which is illustrated in Figure 1. The aim of the study is to assess the magnitude and variation of catch crop yield in relation to location, sowing date and from year to year and to assess the relative importance of various environmental and management factors in determining yield.

Figure 1. : Method for investigating crop yield

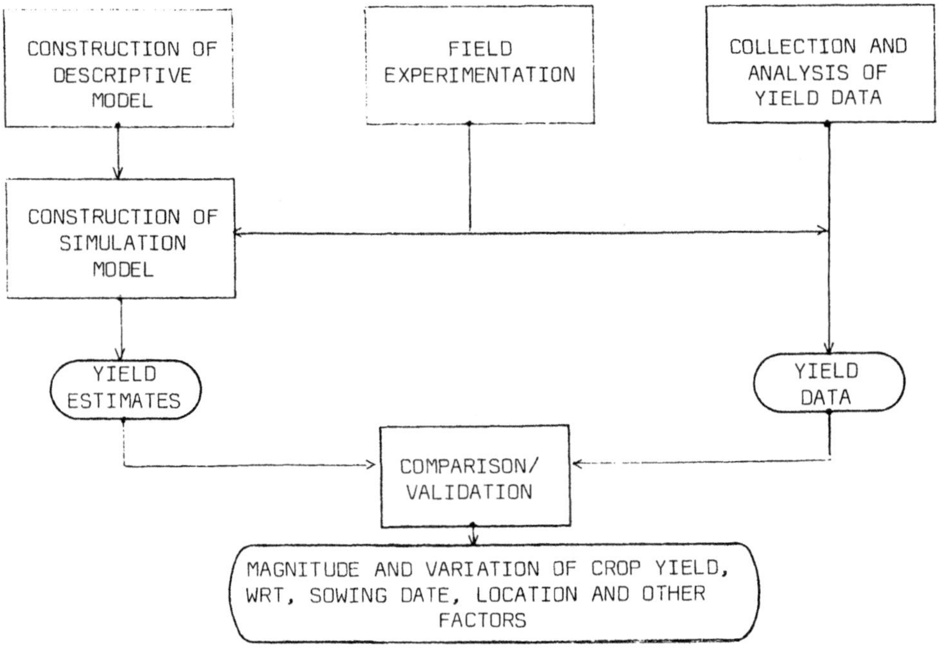

Descriptive Model

The formulation of a descriptive model of the catch fuel crop system is one of the short-term objectives of this project, and is useful in isolating the important yield determinants, in designing field experiments and provides a basis for the simulation model to predict crop yields. One model expression of the catch crop system appears in Figure 2.

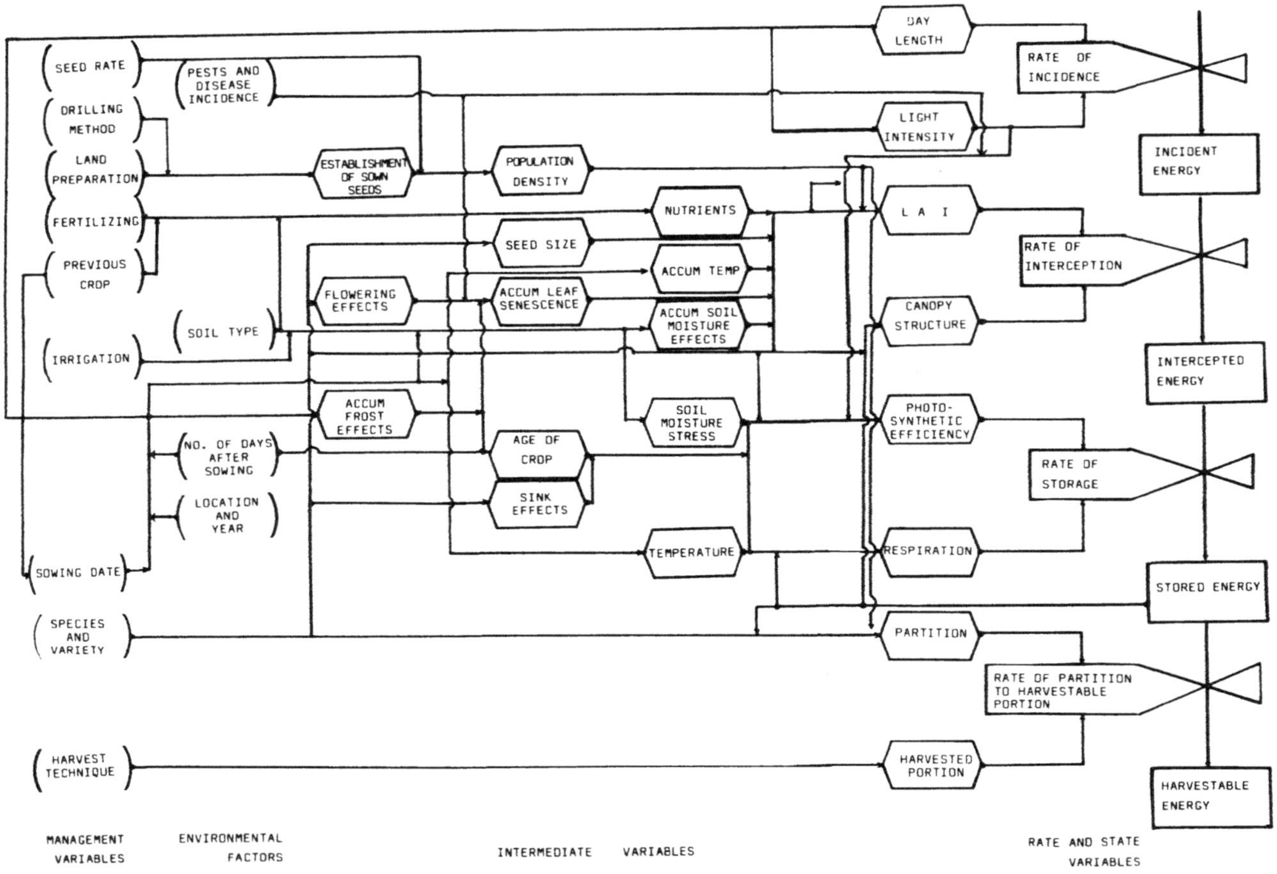

Figure 2 - Relational diagram for the accumulation of harvestable energy by a catch crop

Field Experiment

A field experiment was carried out in the late summer-autumn of 1980 to study the effect of sowing date, species and seed rate on catch crop growth and yield (the second short-term objective of the project) and, hence, provide data for the construction of a simulation model of catch crop growth.

The treatments were as follows:

Sowing dates : 1. July 23
2. August 19
3. September 15

Species : 1. Fodder radish (*Raphanus sativus*) var.'Slobolt'
2. White Mustard (*Sinapis alba*) var. 'English White'
3. Stubble Turnip (*Brassica rapa*) var.'Taronda'
4. Rape (*Brassica napus*) var.'Emerald'

Seed rates : 1. Recommended
2. Five times recommended.

Environmental conditions were kept as optimal as possible; the soil was well fertilized, no drought was experienced during the growing season and weed and pest control were practised where necessary. Periodic growth analyses were carried out and dry matter yields of various fractions, leaf area index, and plant populations measured. Measurements of emergence, ground cover, and light interception were also made, but the instrument used to obtain the latter proved unreliable and a satisfactory set of readings was not obtained.

The experiment was laid out in a split-plot design with two replicates.

3. RESULTS OF FIELD EXPERIMENT

Statistical analyses of above-ground dry matter yield twelve weeks after sowing revealed significant effects only of sowing date ($P<0.01$) and the interaction between seed rate and sowing date ($P<0.05$). At the first sowing date the lower seed rate yielded higher in all four species, at the second only in the faster growing radish and mustard, while at the third sowing date the higher seed rate produced higher yields in all cases (Figure 3). A high seed rate is, therefore, of value particularly where resources are limiting. Similar effects were observed for root and total yield.

Analyses of emergence data suggest that emergence is prompter in the larger seeded fodder radish and mustard and at the earlier sowing dates; although this may account for some of the yield variation between sowing dates it did not effect any significant yield differences between species.

It is hoped to carry out more sophisticated analyses of data from this experiment in combination with results from an experiment to be carried out this year, aimed particularly at demonstrating the relationships between light interception and leaf area index and radiation interception and crop growth rate.

Figure 3. : Table of mean above-ground yield 12 weeks after sowing.

SOWING DATE	SEED RATE \ SPECIES	1	2	3	4	SR X SD MEAN	S D MEAN
1	1	6.52	5.50	5.22	7.81	6.26	5.87
	2	5.65	4.59	5.22	6.45	5.48	
2	1	3.87	4.53	3.91	2.82	3.78	3.84
	2	3.29	4.14	4.19	3.94	3.89	
3	1	0.85	1.07	1.34	0.83	1.02	1.48
	2	1.60	2.30	2.08	1.73	1.93	

4. CONCLUSIONS

The theoretical and experimental studies briefly outlined in this paper have gone towards meeting the intermediate and final objectives of this project. A clearer understanding of the determinants of crop yield has emerged, and data on the effect of yield on some of these has been obtained. Further analyses and experimentation are aimed at the construction of a simulation model to predict catch crop growth on the basis of solar radiation receipt and data from other sources is being collected to aid in the construction and validation of this model.

"JOINT RESEARCH ON ARUNDO DONAX AS AN ENERGY CROP"

Contract Number	:	ESE - 076 - F
Duration	:	36 months from signature of contract
Total Budget	:	I.200.000 F.F. CEC Contribution : 460.000 F.F.
Head of Project	:	M. ARNOUX, Director of the Plant Breeding Station of MONTPELLIER (I.N.R.A.)
Contractors	:	INSTITUT NATIONAL DE LA RECHERCHE AGRONOMIQUE (I.N.R.A.) CENTRE NATIONAL DU MACHINISME AGRICOLE, DU GENIE RURAL, DES EAUX ET DES FORETS (C.E.M.A.G.R.E.F.)
Addresses	:	I.N.R.A. Station d'Amélioration des Plantes 9, place Viala 34060 MONTPELLIER Cedex C.E.M.A.G.R.E.F. 685, route d'Arles B.P. 4020 30000 NIMES

Summary

The wild giant Graminaceous, Arundo donax, is the object of a research programme for its valorization as an energy crop.

The agronomic aspects of the programme are studied by the I.N.R.A. : variability of the species, new management of the crop capable to increase the yield and to reduce the costs of production and the energy input.

The mechanization aspects are studied by the C.E.M.A.G.R.E.F.: adaptation of the harvesting machines to the new methods of cultivation, new technics for the conditionning and the storage of the product, different processes to obtain energy.

Arundo donax, a giant reed of Mediterranean climates, has been studied in the south of France for many years in the aim to use it as a raw material for the production of pulp and paper.

In respect of data obtained, the object of this research pro--gramme is to examine thoroughly the interest of the plant as a new energy crop.

On the occasion of the former meeting in AMSTERDAM, the charac--teristics of the plant and its potential interest as an energy crop have been described. This paper presents a short survey of the pro--gramme and its very first results.

I/ AGRONOMIC RESEARCH

About 40 different types of *Arundo donax* of various origins are observed in order to determine their comportment in south of France and to look for more productive types.

The first observations show that the types of southern origin are the most productive. But because *Arundo donax* is a perennial plant, several years of observation will be necessary to confirm these results.

The culture of *Arundo donax* for the production of pulp and paper was made in fields with a distance of 2 meters between the rows. The average of 10 years was 20 to 25 metric tons of dry matter per hectar and per year, that is 8 to 10 tons of oil equivalent per hectar and per year.

But this form of cultivation involves a lot of concurrency between the rows :

- concurrency for light
- concurrency for water
- concurrency for mineral nutrients
- concurrency for soil occupation by the rhizomes

After 4 to 5 years of cultivation the result is a decrease of the yield year after year.

For these reasons we decided to test a new form of cultivation of *Arundo donax* better adapted to its energetic valorization. That is to cultivate the plant in isolate hedges 15 to 25 meters distant to one another in a classical cereal field.

The results expected are :

- A better use of the land. The wind being particularly violent in this part of France, the hedges will serve as wind-breaks and their protection should have a positive effect on the other crop.
- Lower input of energy and of money. The hedges of *Arundo donax* should only be planted, harvested and their volume controlled every year.
- Equivalent or higher yields. The insolation of the plant and the disponibility of soil for the rhizomes should be at their optimum.

An experimental field has been planted on this pattern this spring and the first results will be available next year.

II/ RESEARCH ON HARVEST MECHANIZATION

Harvesting machines have been perfected a few years ago and are actually working without problem.

The C.E.M.A.G.R.E.F. is in charge of the improvement of these machines and of the building of a new type of machine better adapted to the type of cultivation formerly defined and experimented.

The aim is :

- to simplify the present big harvesting machines which have been conceived to harvest large fields of Arundo donax.
- to reduce their energy consumption.
- to look for and build a smaller machine with a harvesting head capable to harvest isolate hedges, wild hedges and small wild plots of Arundo donax.

III/ RESEARCH ON CONDITIONNING AND STORAGE OF THE HARVEST

I.N.R.A. and C.E.M.A.G.R.E.F. are observing and collaborating to an experiment in a private farm of south of France which has high needs in energy to dehydrate alfalfa, to heat green houses and farm homes and decided to supply these needs with Arundo donax. Researches on conditionning and storage of the harvest and on energy processes are conducted in this farm.

At harvest time the product chopped by the harvesting machine is composed of :

- I5 to 35 % of leaves and sheathes
- 65 to 85 % of stalks

The average water content is about 50 % and the storage of the product just as it is leads to a decomposition of the cellulose by the bacteria.

Researches are actually running on :

- Dehydration of the harvest in the alfalfa dehydration unit in order to lower its water content to I5 % and get a good con--servation of the product.
- Separation and granulation of the leaves in order to reduce the costs of transport and storage and to improve the use of the leaves in a gasification unit.

The first results are very conclusive.

IV/ RESEARCH ON ENERGY PROCESSES

A direct combustion process was first experimented but different problems of temperature, ashes and silicium content of the plant lead us to replace it by a gasification unit which has been settled this spring.

The perfecting of this process is actually in progress with the assistance of the C.E.M.A.G.R.E.F.

Formerly done for the production of pulp and paper, the study of Arundo donax has to be done now for another kind of valorization and requires another point of view.

Different methods of cultivation, new harvesting machines, specific technics of conditionning and storage, convenient energy processes have to be conceived, perfected or tested in order to reduce the costs of production and the energy input.

Arundo donax, a promising new energy crop, has to be considered so at every stage of production, from its planting to its utilization as a vegetal fuel.

SUBJECT II. FORESTRY, WOOD, WASTES

An experimental study of short rotation forestry for energy

Coppiced trees as energy crops

Selection and creation of forest material for biomass production
Establishment of short rotation coppices

Utilization of coppice forests biomass for fuel and other industrial uses

The production of energy from short rotation forestry

To design, build and test a biomass harvester/chipper

Design and building of a forestry wastes harvester

AN EXPERIMENTAL STUDY OF SHORT ROTATION FORESTRY FOR ENERGY

Contract Number	:	ES-E-R-016-UK(N)	
Duration	:	36 months	1 July 1980 - 30 June 1983
Total Budget	:	£112,000	CEC contribution £50,400
Head of Project	:	Professor J D Matthews	
Contractor	:	Aberdeen University, Forestry Department	
Address	:	Aberdeen University Forestry Department St Machar Drive ABERDEEN AB9 2UU UK	

Summary

The work done in the first year of a contract with the CEC is reported. The programme has three sections; assessment of above-ground biomass produced by forest trees, establishment of trials of short rotation forest energy plantations, and preparation of specifications for equipment for harvesting forest biomass for energy.

The first section is complete. The dry matter production of ten tree species has been examined across a range of ages and yield classes for total biomass and partitioning between stem, branches and foliage. Mathematical relations between stand dry matter production and mensurational parameters have been developed. A set of tables (for conifers) which can be used to predict dry matter production from crop basal area is now available.

Three trial short rotation forest energy plantations were planted with the same ten tree species in the spring 1981. These were on marginal farmland, scrub woodland and the site of a young plantation in the north east of Scotland.

A survey of harvesting equipment and systems is in progress. The development of a tractor-mounted machine for harvesting coppice for energy is proceeding under a sub-contract.

The three year programme is in three sections:

- determination of production of biomass above-ground from young plantations of single-stemmed trees,
- establishment of trials of short rotation forest energy plantations in typical locations in Britain,
- preparation of specifications for equipment and systems suitable for harvesting forest biomass for energy.

Section 1 : Production of Biomass

The main aim is to provide estimates of dry matter production above-ground and its partitioning within trees for ten tree species aged 12 to 28 years.

The objects were:

- to tabulate, for each site, total above-ground biomass per tree and its partitioning between stem, branches and foliage,
- to derive mathematical relations between total biomass and its components and some easily measured tree parameter for single trees (the Single Tree Model),
- to derive a mathematical relation between stand dry matter production and some stand mensurational parameter (the Stand Model),
- to produce a set of stand biomass tables to aid management of forests for energy.

The ten tree species were alder, birch, *Nothofagus*, sycamore, Corsican pine, Douglas fir, hybrid larch, Scots pine, Sitka spruce and Western hemlock. These were selected for their potential as producers of biomass and were assessed at ages up to 28 years and at several growth rates for total biomass production above-ground and its partition between stem, branches and foliage. The effect of geographic variation in site quality on biomass production was examined by stratifying the sample by Yield Class which indicates site production potential. Twenty-seven sites were sampled.

At each site the age, basal area at breast height, tree stocking and hence Yield Class were determined. Twelve trees were sampled destructively to determine total dry weight above-ground and its partitioning between stem, branches and foliage.

Single tree models as linear regression equations of total biomass and its partitioning per tree on basal area per tree at breast height were determined for each site. These were used to construct single tree tables. Values of basal area at breast height were computed for one centimeter stem diameter classes in the range of diameters sampled and these were substituted in the component regression equations to produce the single tree tables.

The predicted biomass yields for each site were computed by substituting the mean basal area of the 100 trees measured in each sample stand in the appropriate single tree models and multiplying by the stocking. The predicted total biomass yields are given in Table 1.

Table 1 PREDICTED YIELDS OF BIOMASS

Species	Age (yrs)	Yield Class (M^3/yr)	Standing Dry WT (t/ha)	Productivity (t/ha/yr)
Alder	12	6	26	2.2
	18	10	74	4.1
	22	14	210	9.6
Birch	25	6	100	4.0
	20	9	99	5.0
Nothofagus	13	17	75	5.6
	24	11	145	6.0
	17	14	102	5.9
Sycamore	28	4	97	3.5
	22	10	114	5.2
Corsican Pine	19	12	86	4.5
	20	16	136	6.8
Douglas Fir	14	20	86	6.2
	18	12	68	3.8
	21	16	126	6.0
Hybrid Larch	13	10	36	2.8
	20	10	111	5.6
	20	15	156	7.8
Scots Pine	14	14	43	3.1
	18	8	60	3.3
	19	14	109	5.7
Sitka Spruce	14	26	109	7.8
	22	14	105	4.8
	20	26	209	10.5
Western Hemlock	15	28	174	11.6
	17	16	58	3.4
	18	20	109	6.1

Table 2 STAND BIOMASS TABLE FOR CORSICAN AND SCOTS PINE

For Crops: Age 14 to 20 years

Yield Class 8 to 16

Stocking 2800 to 4405 stems/ha

Basal Area 24.43 to 48.88 sq m/ha

Basal Area m^2	Total Dry Weight t	Tolerance %
25	49	32
26	55	24
27	61	20
28	67	17
29	72	16
30	77	16
31	82	16
32	87	16
33	91	16
34	96	16
35	100	16
36	104	16
37	108	15
38	111	15
39	114	14
40	118	14
41	120	13
42	123	13
43	126	12
44	128	12
45	130	12
46	132	13
47	134	13
48	135	14

Stand models were derived for four populations from the single tree models:- the pine to 5 and 7 cm top diameter and the remaining conifers to 5 and 7 cm top diameter. Hardwoods were excluded because there were insufficient data to cover the total range in basal area.

These stand models are multiple regression equations of total dry weight per hectare on basal area per hectare and basal area per hectare squared.

Stand tables for the four populations covering the range of basal areas within each were produced for one square meter basal area classes. These can be used by forest managers to determine yields of forest biomass for energy from even-aged pine plantations. The information required is:- species, age, Yield Class, stocking and basal area which are used in the management of plantations for timber production. Once the basal area of the crop is known the relevant table can be consulted and predicted biomass yield obtained.

A stand table which can be used to predict total standing dry weight from basal area per hectare for Scots and Corsican pine appears in Table 2.

Section 2 : Establishment of Trials of Short Rotation Forest Energy Plantations

The object is to define the economics and logistics of establishing forest energy plantations. The plantations are being established on three site types; marginal agricultural land, existing scrub woodland and to replace young plantations. The latter simulates the condition of successive crops of biomass plantations. The ten selected species listed in Section 1 are being tested in comparative replicated trials situated in four geographic regions of Britain. These are north-east Scotland, the borders between England and Scotland, south-west England and the borders between England and Wales.

The three site types in north-east Scotland were planted in spring 1981. One site is an area of marginal agricultural land at the boundary between two good-quality fields. The site was fenced and not ploughed. Planting with bare-rooted stock was done at the end of March 1981. Intensive weeding will be done during the first growing season.

The scrub woodland site had a cover of birch which was removed and sold for firewood. The site was ploughed to control future weed growth and aid establishment. Planting was done in May 1981.

The young plantation site carried Scots pine with some larch and Douglas fir. This crop was sold standing for pulp wood and pallet wood. Following harvesting of stems by contractors the remaining tops and branches were cleared by machine. The site was planted in April 1981.

Detailed costings for all the operations are being recorded. The remaining sites will be planted in spring 1982.

Section 3 : Harvesting Forest Biomass for Energy

Forest biomass for energy will be available from the following sources:-

- clearing scrub woodland;
- forest residues;
- short rotation single stem forest energy plantations;
- short rotation coppice forest energy plantations.

A survey of machines and systems for harvesting single stem trees grown on short rotation and harvesting forest residues for energy is being undertaken.

The development of a tractor-mounted machine for harvesting coppice hardwoods grown on disadvantaged farmland has been sub-contracted to Loughry College, Northern Ireland. The first prototype has been built and was tested in 1981 on a 2 year old willow crop and a simulated crop. The trials were encouraging but problems were encountered; there were more small diameter willow stems in the crop than anticipated. This caused problems in removing the cut stems from the saw blade to the packing device.

Modifications to the machine are being made and it will be tested later this year.

COPPICED TREES AS ENERGY CROPS

Author	:	M.L. PEARCE	
Contract number	:	ESE-R-017-UK	
Duration	:	36 months	July 1980 - June 1983
Total Budget	:	£52,000	CEC Contribution : 40,000 U.A. (£24,400)
Head of Project	:	D. R. Johnston, Director Research	
Researcher	:	M. L. Pearce, Research Chief Forester	
Contractor:	:	Forestry Commission - Research and Development Division.	
Address	:	Westonbirt Arboretum Tetbury Gloucestershire England	

Summary

This paper reports progress in the first year of the Contract. Various difficulties have prevented the establishment of four of the seven proposed experiments, but these will be completed during the winter and spring of 1981/82.

1. Contract

(a) To establish seven field experiments, designed to define production models for a range of tree species and site types; using a coppice system of energy feedstock production.

(b) Assess costs (£ per dry tonne of feedstock at the forest or farm gate) to form the basis of an economic evaluation.

(c) Make available the material harvested in these production experiments, for the evaluation of:

(i) harvesting equipment.

(ii) feedstock conversion systems to utilisable energy.

(these latter evaluations are the subject of other contracts)

1.1. Experiment Plan(Model)

7 sites selected from 4 broad environmental zones;

(i) High rainfall - low irradiation.

(ii) Low rainfall - high irradiation.

(iii) Moderate rainfall - high irradiation.

(iv) Hot-dry tertiary soils.

and 2 land class categories;

(a) Marginal agricultural - high fertility, small scale, and under utilised.

(b) Low grade broadleaved woodland.

3 experimental treatments per site;

3 Species (from *Populus*, *Salix*, *Nothofagus*, *Alnus* and *Eucalyptus*).

X

2 Stool spacings (10000 ha^{-1} and 2500 ha^{-1}

X

2 Harvesting rotations (cut every other year and cut every 4th year).

Sufficient replication of all treatments on each site, to provide a sound statistical model. (3 replicates).

1.2. Progress. The original intention was to establish experiments on 7 sites by spring 1981. Unfortunately, difficulties over securing sites, preparing them, and obtaining the right plants has prevented achievement of this.Inclement weather during the last winter and spring reduced the opportunity for much of the establishment work. Together these factors resulted in only three of the seven sites being planted. A major factor in the delays has been the time of the researcher allocated in the contract - this has proved inadequate to cover the work necessary at all seven of the geographically widely spread out sites. Some additional assistance will be sought in order to complete the establishment phase during the winter-spring 1981/82.

Table I

Land Use / Environment	Marginal Agriculture	Low Grade Broadleaved Woodland
High rainfall Low radiation	Shropshire Establish winter 1981/82	Shropshire Establish winter 1981/82
Low rainfall High radiation	Mepal - Cambridgeshire. Experimental Husbandry Farm. Established spring 1981. Poplar, willow, *Nothofagus* (*Eucalyptus*).	
Moderate rainfall High radiation	Bristol - Avon Long Ashton Research Station. Established spring 1981. Poplar, willow, *Eucalyptus* (*Nothofagus*).	Somerset/Wiltshire Establish winter 1981/82
Hot-dry tertiary soils	Establish winter 1981/82	East Sussex Manley Wood Established winter 1980/81 Poplar, willow, natural coppice.

1.3 Prognosis

It is far too early to assess any results from this project, but some observations can be made.

(i) The production curves projected for a coppice system (figure I) are steadily being substantiated by many researchers in other parts of the world - recent work on poplar in Italy is particularly encouraging (1).

It is important to recognise that maximum mean annual increment (which determines rotation length) will vary with species and site and at least 10 years will be required to obtain the optimum result from this project.

(ii) It is becoming increasingly evident that maximum production will only be achieved when weed-free conditions exist. More intensive herbicide weed control will create a greater financial input than was originally estimated.

(iii) The choice of species within the candidate Genera is being made arbitrarily, ie. which species of *Nothofagus*?, which clone of poplar? Further research into these variations, especially with *Alnus* and *Salix* is suggested by the International Energy Agency, and should be supported.

1.4. Finance

A review of the financial estimates is desirable and will be considered by the main contributor - the U.K. Department of Energy.

Additional support staff (para. 1.2.), additional materials (para. 1.3 (ii)), and inflation, are put forward for consideration as follows:-

E1.1.	Staff expenses	- Inflation (7%)	£ 2,500
	" "	- Additional support (1 yr.)	£ 6,000
E1.5.	Other expenses	- Inflation (10%)	£ 900
	" "	- Additional chemical	£ 600
		Total -	£10,000

This would increase the total contract to £62,000, with a possible additional contribution from C.E.C. of £4,000.

2. Conclusions

The delay in the programme, although regrettable, will not seriously affect the analysis of the results - any environmental variations between planting dates can be assimulated into the production model. Obviously, the total length of the project has been extended by one year, but perhaps this is less significant if measured against the 10 years projected in para. 1.3 (i). In all other respects the project is on course, but it is far too early to give any results.

References

1. Rossi Marcelli et al. 1980. 'The productivity of a poplar plantation'. Cent. Sper. Agric. For. Rome. 13: No. 3. 203-227.

YIELD CURVES

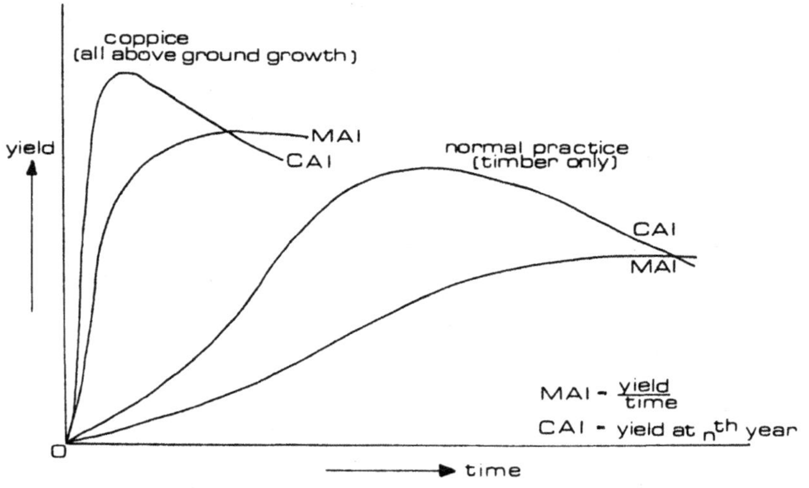

Figure I

Figure 1

1 - SELECTION AND CREATION OF FOREST MATERIAL FOR BIOMASS PRODUCTION

2 - ESTABLISHMENT OF SHORT ROTATION COPPICES

Contractor number : ESE-R-018-F

Duration : 36 months 1 july 1980 - 30 june 1983

Total Budget : FF 1.000.000 CEC Contribution : FF 46.000
not available june 1981

Head of Project : J.F. LACAZE, Institut National de la Recherche Agronomique
Station d'Amélioration des Arbres Forestiers

Program run by : Eric TEISSIER du CROS, same institute, same laboratory

Address : Ardon
45160 OLIVET
FRANCE

Summary

A particular selection and improvement program has been started. Its objectives are to widen the amount of tree species to involve in short rotation coppices devoted to biomass production, and to study the behaviour of these species under possible conditions of sylviculture adapted to this production and to the site conditions which will probably be reserved for it. The tree species will be Alders, Poplars and Aspens in a first turn to be planted in 1982 and American Red Oak, Black Locust and Sitka Spruce in a second one to be planted in 1983. Factors of variation of the environment and of the sylviculture will be : quality of soils, spacings, rotations and probably fertilization and Frankia inoculations (for Alders). Considering the youth of these trials, no results are yet available. First results will hopefully occur at least 5 years after plantation.

1 - INTRODUCTION

Considering the permanently increasing price and possible lack of fossil fuels, France, following in that many other countries, has decided to cover part of its needs in energy with all the possible other sources. Amongst these is the conversion of solar energy into biomass.

Forest, although not the most efficient route to convert solar energy into biomass, is probably the nearest one to feasability and practice. Unfortunately it has been considered in France that a large increase of the afforested area would be quite difficult. As a matter of fact only small areas of farm land will be devoted to afforestation and forest land will still have to produce large quantities of sowlog and industrial wood for which the commercial balance gets more unfavorable from year to year. Only marginal sites are left.

Two methods are possible to produce forest biomass :

1 . Evaluate and cut the existing coppices, try and increase their productivity with proper sylvicultural methods

2 . Create new stands devoted to produce biomass.

The first method is already studied in Italy (see corresponding paper in the same proceedings) and in France by the laboratory of Sylviculture and Environment in Orleans (INRA).

We shall here consider the second one.

2 - OBJECTIVES

Lots of information is available from trials made in France (AFOCEL), western and northern Europe (The Netherlands, Great Britain, Sweden) and North America. Yields vary within large limits but concern a small number of species(mainly poplars and willows) and sites (mostly abandonned farm land and fairly good soils).

As said before, France will probably have to concentrate its biomass studies and production in conditions marginal both for agriculture and forestry. The main characteristic of these marginal conditions is hydromorphy. Most of our trials will thus be planted in such conditions.

Another objective of the studies we have started is to widen the number of potential species envolved in biomass production.

It is therefore considered that a proper and powerful selection and improvement program has to be started. We shall now give a list of the species which will be studied and the different factors which they will be confronted with.

3 - FOREST TREE MATERIAL

The first year of the contract has been devoted to the collection of forest tree material and the beginning of its education in our nursery at Orleans.

Alders	:	Black Alder	*Alnus glutinosa*
		Grey Alder	*A. incana*
		Corsican Alder	*A. cordata*
		Red Alder	*A. rubra*
		Japenese Alder	*A. hirsuta* *A. inokumae*

Poplars :	Black Cottonwood	*Populus trichocarpa*
	Grey Poplar	*P. canescens*
	Hybrid Aspen	*P. tremula* x *tremuloïdes*
Black Locust		*Robinia pseudo-acacia*
American Red Oak		*Quercus borealis*
Sitka Spruce		*Picea sitchensis*

Although certain alders when used as pure species, seem quite promising we think necessary to look for still more productive material within this Genus. Interspecific hybrids come to mind. A certain amount of crosses have been tried in spring 1981 (see table I).

Table I - Intra- and interspecific crosses of Alder performed (X) in spring 1981 at Orleans.

Female species \ Male species	*Alnus glutinosa*	*Alnus cordata*	*Alnus rubra*	*Alnus hirsuta*
A. glutinosa	x	x	x	x
A. cordata	x	x	x	
A. incana	x	x		
A. hirsuta	x	x		
A. inokumae	x			

4 - SITE AND SYLVICULTURAL CONDITIONS OF THE TRIAL

Although we have not planted the first biomass trial yet, we have had to think upon the different factors of variation to which the different species have to be submitted.

4.1 Sites - They are mostly hydromorphic
- Silt layer on a pseudogley horizon for Black Cottonwood
- Sand layer of varying depths on a pseudogley horizon for Alders.

4.2 Spacings - In a first series of tests to be planted in spring 1981, in which the main purpose is to estimate the adaptability of the forest tree materials, plantation density (being a secondary characteristic), will only have two levels : 2000 and 4000 plants per hectare.

4.3 Rotations - For the same reasons we shall only apply two periods of rotations. Date of coppicing will be determined from annual increment measurements and occurence of competition effects.

4.4 Fertilization - Although it might seem queer to imagine an investment of fertilizers which in most cases need energy to be produced and transported, it has been shown that fertilizers applied at plantation produce a starter effect which gives a favorable economical balance at harvesting when compared to unfertilized controls. This factor might be considered in our tests in a second series to be planted in 1983, but advices are asked for in this field from other participants to this contract.

4.5 Inoculation of Frankia - The Alder-Frankia symbiosis for Nitrogen fixation on root nodules is well known. It is also clear that nodules appear on roots as soon as Alder seed is sown and wherever it is sown. But the question is raised whether the existing Frankia in the soil will enable the highest production of Alders used. This question is still more important if one thinks of possibilities of introduction of exotic Alders and of interspecific hybrids. This interaction is already under study in different part of the World (UK, The Netherlands, USA, Canada). Strains of Frankia have been isolated and cultivated by Dr LALONDE in QUEBEC and by a laboratory in Lyon (France). A preliminary test comparing the effect of inoculation of Frankia in a seed bed compared to no inoculation on seed lots of *Alnus glutinosa, A. rubra, A. cordata* has been established at Orleans in June 1981 in cooperation with Rhône-Poulenc research department. Other semilar tests will be established in the future with a wider range of Frankia strains and Alder species in cooperation with the Research Laboratory for Soils and Fertilization of the Forest Research Institute at Nancy (France).

5 - CONCLUSION

All the studies we have started are yet to young to have produced any results. Within a few months we shall hopefully get an answer to two questions raised :

1 . possibility to get seed from a series of interspecific hybrids

2 . competitivity and effeciency of a strain of Frankia inoculated in the sowing medium of Alder seeds.

For the main problems considered, i.e. possibility to select tree species for biomass production and possibility to produce a sufficiently large quantity of biomass in marginal forest soils, we shall have to wait at least five years after plantation. This leads us to 1987 for theme 1 and 1988 for theme 2, at the earliest.

UTILIZATION OF COPPICE FORESTS BIOMASS FOR FUEL AND OTHER INDUSTRIAL USES

Contract Number	:	ESE-R-035-I	
Duration	:	36 months	1 July 1980 - 30 June 1983
Total Budget	:	L. 495,400,000	CEC Contribution :
Head of Project	:	Prof. G. Scaramuzzi, SAF-Centro Sperimentazione Agricola e Forestale, Roma	
Contractor	:	Ente Nazionale Cellulosa e Carta	
Address	:	V.le Regina Margherita, 262 00198 ROMA	

Summary

The project is part of a larger one promoted by the National Agency for Cellulose and Paper (E.N.C.C.) for an enhanced utilization of coppice forests in Italy, and concerns an experimental area in Calabria. The research programme covers silvicultural, harvesting and technological aspects. Investigations and trials have started only recently; therefore, an outline of the research expected to be carried out is given, together with a short description of the experimental area.

1. Introduction

A description of Italian coppice forests and of the relative silvicultural and utilization problems has been given in the report submitted to the preceding coordination meeting in Amsterdam.

In view of the ever more increasing difficulties met by the paper industry for raw material supply and in the belief of contributing to the solution of one of the most prominent problems of Italian forestry, the National Agency for Cellulose and Paper (E.N.C.C.) has promoted a large research project for an enhanced utilization of coppices in Italy, to be carried out by its scientific and technical branches in cooperation with other research organizations interested in the problem.

The research programme concerns the coppices for which no conversion is expectable. Its object is to find out technically and economically suitable harvesting methods in agreement with the safeguard of forest productivity, and to assess the possible suitable uses of the obtainable biomass.

The trials are expected to cover different areas representative of the main Italian coppice types, and various geomorphological characteristics.

The project has recently been granted a financial support by the Commission of European Communities.

2. Outline of the contract-project

The contract-project is part of the larger ENCC's project and concerns an experimental area in Calabria. It will cover the following aspects:

- silvicultural aspects, concerning assessments and investigations dealing with site and stand characterization; forest production; standards selection in relation to the harvesting system used; machinery effects upon stumps and standards vegetative conditions, natural regeneration, stability and evolution of soil and vegetation;
- harvesting aspects, including felling, skidding and chipping trials of the obtainable stock;
- technological aspects, including trials and estimates on the use of coppice biomass for energy, on the ground of its technological properties and of limitations for its industrial use.

3. Preliminary results and further investigation expected

Soil and coppice stand preliminary assessments as also utilization trials were to be ended by July 1981, compatibly with the climatic trend. The aspects concerning the energetic use of coppice biomass and short-term observations on mechanization effects will be ready by the contract expiration.

Because of an unexpected delay in the acquisition of the experimental area and of adverse climatic conditions over the first months of the year, it was possible to start the stand and site investigation only in April, and the harvesting trials in late May.

The selected experimental area, located in the district of Cerva, Catanzaro (Calabria), within the Regional Forest Property, is representative of oak formations, with a prevalence of Quercus cerris, Q.farnetto and Q.pubescens, covering very large surfaces in southern Italy, for part of which the conversion into high forest - even if feasible - does not prove convenient. In the case of the area under study, its non-convenience mostly depends on the stand stucture (presence of a very large number of small- and very small-sized sprouts), which would bring about conversion costs so high as to find in any case no counterbalance from a hypothetical increase of value of the resulting high stand.

The area is located at 800-1100 m a.s.l. with a south-west prevailing exposure and a gradient exceeding 30% (up to 100%). The forest cover is chiefly represented by Q.cerris.

The site and stand investigation is quite completed, the following assessments being carried out: stand age, stem and stump density, stand production, current and mean increment, mean and species green and basic wood density, mean and species bark percent, pedological characteristics.

The harvesting trials are underway. A first lot of some 8 hectares is expected to be utilized by June, while the remaining area will be cut by the end of autumn.

The complete-chipping system is being adopted, with the following scheme of work:

- felling by motor-saw and piling up of whole sprouts;
- concentration by skidder (on short distances) or by skyline (on long distances);
- hauling by skidder;
- chipping of whole sprouts at the landing, and loading on lorries.

Times, yield and costs of the various working phases are also being assessed.

After-felling observations will be carried out to qualitatively and quantitatively evaluate the eventual damage caused by mechanical operations to stumps, standards and soil. Short- and mean-term observations will be also conducted on machinery effects upon the resumption of vegetative activity, on wood increments, soil and vegetation stability and evolution.

Trials on the industrial use (fibre boards and paper pulps) of the obtained coppice biomass will be performed within the framework of the larger ENCC's project. Moreover, the following trials and estimates on coppice biomass use as fuel will be made:

- the calorific value of the material from the different species constituting the main types of Italian coppices;
- compatibly with CNEN's availability of suitable burners, combustion trials in cooperation with the same organization;

- the expectable availability of coppice biomass for energy uses, on the ground of its technological properties as well as of limitations for its industrial use, together with an analysis of the conditions required for its economic utilization as an energy source.

THE PRODUCTION OF ENERGY FROM SHORT ROTATION FORESTRY

Contract Number	:	ESE-R-036-EIR H	
Duration	:	36 months	1 July 1980 - 30 June 1983
Total Budget	:	£Ir 398,177	CEC Contribution : £Ir 173,800
Head of Project	:	Dr. M. Neenan, Agricultural Institute, Oak Park	
Contractor:	:	An Foras Taluntais (translated - Agricultural Research Council)	
Address	:	19 Sandymount Avenue Ballsbridge Dublin 4.	

Summary

The objective was the production of woody biomass on a 3 to 5 year cycle as a feedstock for power stations. In the spring of 1977 field trials were laid down at 5 centres. Since then more intensive studies have been made on two peatland areas. Preliminary yield data obtained by the end of 1980 have been subjected to a systems analysis. Present indications are that, excluding land prices, short rotation forestry fuel can be produced at a price of about $104 U.S. per TOE, with a break even point of 5 to 16 years, depending on the inputs. It is foreseen that these costs can be further reduced by the use of improved clones, less expensive soil treatments and other technologies now being developed.

In utilisation, the main problem has been the relatively flat shape and high moisture of the particles. Studies on the rate of drying in the open air are in progress, and close liaison is being maintained with the 5MW combustion trials of the DGXVII project.

There is considerable interest in the project as a potential source of heat for private dwellings and small rural industries.

1.1 Introduction

The object is to produce short rotation forestry for direct combustion at minimum cost. In Ireland there are six types of land which are potentially available.

(a) Mountain land (podsols) - Kilfinane

(b) Wet hilly abandoned farmland (gleys) - Swanlinbar

(c) Second quality agricultural land - Oak Park

(d) Blanket peat: virgin d_1 - Ross
after harvesting for fuel d_2 - Glenamoy

(e) Fen peat: virgin e_1 - Derrycoffey
after harvesting for fuel e_2 - Clonsast

The main task of the investigation is to determine the species and the cultural methods by which this fuel can be produced at a price which is competitive with oil.

1.2 Methodology

In the spring of 1977 field experiments were laid down on the 5 soil types (see Fig.1), those on peat being on virgin bog only. Trials (a),(b), (d_1) and (e_1) were subcontracted to the (State) Forest & Wildlife Service (FWS). From a study of the literature (1), and the practical experience of the FWS, the list of candidate species/clones was reduced to 12. In 1978 the trials were extended using a slightly different range of species. At the harvested fen peat centre (Clonsast) a number of aspects such as fertiliser application, spacing, effect of time of harvesting etc. was investigated in more detail.

Studies were made on cultivation, harvesting and transportation costs on the basis of published data and contractors prices. Trials on the combustion properties of different types of chipped wood showed the main problems as particle size, particle shape and excessive moisture content. A series of storage/drying experiments has since been established.

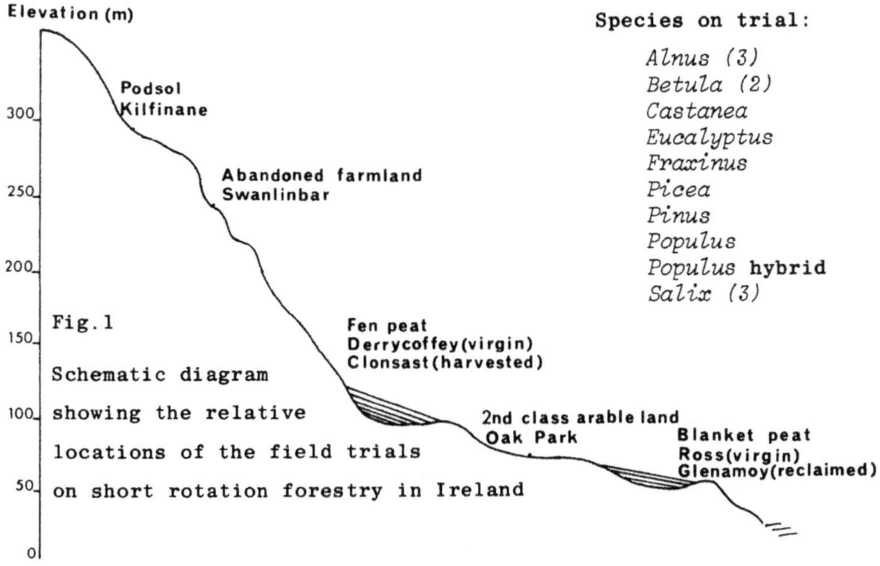

Fig.1 Schematic diagram showing the relative locations of the field trials on short rotation forestry in Ireland

The cost effectiveness of the entire operation is being evaluated by the programmes SALIX and IMPACT. The former has been developed by simulation mode but by now the first of the physical data are becoming available.

1.3 Results

In category (a) soils, which cover approximately 22% of the area of Ireland, it is possible to grow conifers with moderate success from rooted plants. With further research, such areas may be capable of growing conifers of a quality suitable for fuel. Rotation time, initial plant spacing, and ways of reducing cultivation costs at establishment require further investigation.

Good results were obtained on previously cultivated or otherwise well prepared land (soil types (b),(c) and (e2). Estimates based on 3 plants of two poplar clones at Oak Park, have shown that the coppice yield is substantially better than the seedling yield, and that the advantage of 0.3m x 1m over 0.6m x 1m spacing will have disappeared by the first coppicing. The growth of poplar at Swanlinbar was excellent, but the plant population (1.2x1.2m) was too low to give optimum results at this stage. Some other species are also promising at this site.

Virgin peat is not now likely to become available in quantity as all such areas will be harvested for fuel where possible. It is foreseen however that some pockets of peat will not be economic to extract. Energy plantations for electricity production are therefore likely to be spread over a patchwork of different soils, each having a species of tree which is best adapted to that area. On the two sites on virgin peat, only conifers are successful up to now. On the reclaimed blanket peat at Glenamoy, good establishment was achieved with rooted plants of *Salix aquatica gigantea*. Some *Alnus* plants from an older plantation were found to be well nodulated with indigenous *Franki* spp. On the harvested fen peat at Clonsast, *Salix aquatica gigantea* gave a yield of 15 t ha annum from cuttings.

TABLE 1.Economic analysis of short rotation forestry for different centres using the SALIX and IMPACT programmes (The discount rate is 6%)

Activity	Frequency of occurrence	Cost at centre (Ir£/ha)		
		Oak Park	Clonsast	Swanlinbar
Land purchase	Establishment	6175.00	500.00	1112.00
Site preparation	"	37.00	37.00	50.00
Planting	"	1558.30	2400.83	1041.60
Fertilisation	"/each harvest	52.70	94.50	Nil
Maintenance	Annually	23.00	23.00	23.00
Harvesting	Each harvest	4.50 per green tonne		
Transportation	" "	2.10 per green tonne		
Yield DM t ha year		10.0	15.13	2.80
Best species		*Populus trichocarpa*	*Salix a.gigantea*	*Populus trichocarpa*
S.R.F.production cost Ir£/TOE:		184.73	78.3	179.06
Benefit-cost ratio:		1.14	5.66	1.17
Break-even period:		28 years	5 years	28 years
Excluding land cost-				
S.R.F. production cost Ir£/TOE:		68.90	67.18	113.56
Price U.S. dollars per TOE:		106.8	104.1	176.0
Benefit-cost ratio:		3.05	6.60	1.85
Break-even period:		8 years	5 years	16 years

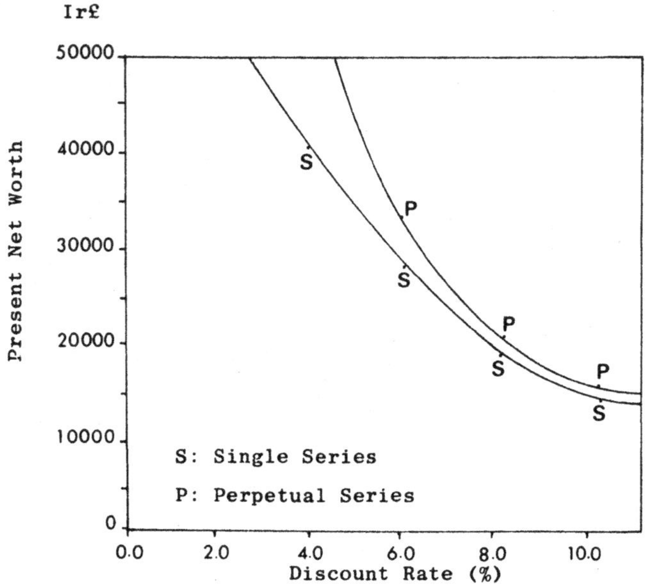

Fig.2 Variation of Present Net Worth with Discount Rate on the harvested fen peat site at Clonsast.
Species: *Salix aquatica gigantea* established from cuttings

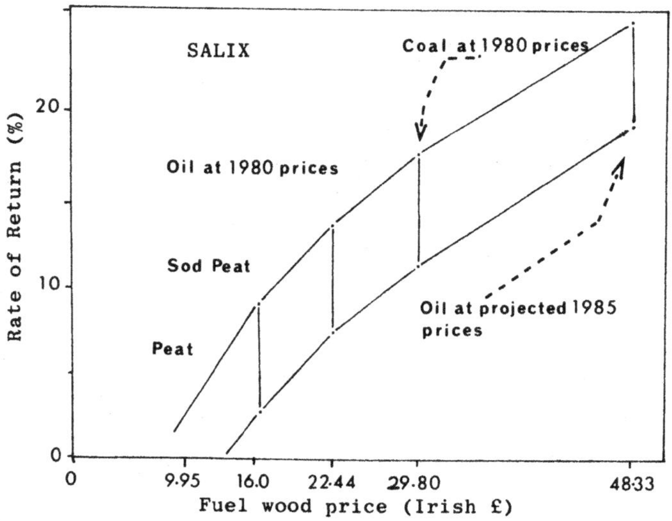

Fig.3 Variation of rate of return on fuel wood plantations when compared to other sources of energy. Range represents 2-year to 5-year curring cycles

Since it would be pointless to give a lengthy account of the negative results, this report is confined to applying the statistical analysis to the best species or treatment on each type of soil.

1.4 Analysis of results and comments

The lowest cost scenario, namely establishment of plantations of willow and poplar by means of cuttings, is only possible at centres (b), (c) and (e_2). The economic analysis for these three scenarios is given in Table 1. This shows that it is possible to produce fuel at a price equivalent to 104 U.S. dollars per ton of oil equivalent in June 1981, assuming that the cost of the land can be discounted over an infinitely long period. If current prices of land are taken into account, the cost of production is 121 to 285 U.S. dollars per TOE. The higher figure is close to the present world price of oil. Projections of this type can be criticised on the basis that the discount rate used was arbitrary. Fig.2 shows the sensitivity of costs to interest rates. This is appreciable at low rates but tends to level off at about 8%. The use of 6% therefore represents a fair compromise within the range which is likely to occur. The economic rate of return on the plantations as compared to other fuels is shown in Fig.3. On the basis of calorific values short rotation forestry would give an annual return of about 13% at oil prices and about 15% at coal prices for the year 1980.

2.1 Conclusions

It must be emphasised that the results as reported here are based on very limited yield data. The yields obtained are however in agreement with those from N.Ireland (2). The increased yield of coppice over primary growth is similar to that being obtained in the U.S.A.(3).

The number of clones tested and the duration of the trials make it unlikely that the best combination of soils and species has been determined. There are now indications that poplar can be grown at Clonsast, and a number of new clones have been obtained from Belgium, Germany, and the U.S.A. with a view to further field trials in 1982. It is reasonably certain that the cost of establishment can be reduced through the adoption of less expensive methods of land preparation and drainage. If the *Alnus* coppice should prove reasonably productive on the peatland, there would be a saving in fertiliser application.

The technology of short rotation forestry for fuel production is at an early stage of development, and with further research it should be possible to reduce costs well below the present equivalent price of oil. While the project was initiated for the fueling of power stations, there is now a considerable interest in short rotation forestry for domestic fuel and for process heat in small rural industries.

References

(1) Neenan M. and G. Lyons. Production of Energy from Short Rotation Forestry. An Foras Taluntais, Dublin. EEC Biomass contracts 405-77-12 ES-EIR and E-E-R 007 EIR N 1980.

(2) Anon. Annual Report 1979. The Horticultural Centre, Loughgall, Dept. of Agriculture, Northern Ireland, p.56-59, 1979.

(3) Dutrow G.F. Economic implications of silage sycamore. USDA For.Serv. Res. Paper SO-60 Southern For.Expt.Stn. New Orleans, 1971.

TO DESIGN, BUILD AND TEST A BIOMASS HARVESTER/CHIPPER

Contract No	:	ESE-R-019 O EIR
Duration	:	36 months 1/7/80 - 30/6/83
Total Budget	:	IR£278,350 C.E.C. Contribution IR £80,720
Head of Project	:	J. Healy, Bord na Mona.
Contractor	:	Bord na Mona Lower Baggot Street, Dublin 2, Ireland.
Sub-contractor	:	Irish Sugar Co. Carlow, Ireland.

Summary:

The objective of the project is to design, build and test a harvesting system for short rotation forestry. The system to be capable of cutting and processing a wide range and type of trees, work on a wide range of terrains from peatlands to sloping ground, be flexible and mobile.

A design team has been set up drawing experience from agricultural machinery development, peat production and forest harvesting.

Machine specifications have been drawn up and a final design agreed on.

The designed machine will cut the material at ground level and convey it to a billeting mechanism. Here it will be processed into approximately 13 cm lengths with a maximum of 8 cm diameter. Following this it will be further processed by a flail device and conveyed to a hopper.

Construction of this machine has now commenced.

Objective:

The objective is to design, build and field test a harvesting system for short rotation forestry. The device must cut and process a wide range of trees of various types and sizes. It should be capable of traversing a wide range of terrains from peatlands to sloping ground. The machine should be flexible and mobile. It should leave the plantation in a suitable condition to facilitate coppicing and regrowth of the cut stumps while at the same time collecting all grown material without waste. The system to have minimum energy requirements and low labour demand. The material produced should be in such a form as to take maximum advantage of solar drying with minimum rewetting. The material should also be capable of storage over long periods, without deterioration. It should also be in a form to allow for mechanical handling, ease of transport and efficient conversion with, if possible, no further processing or treatment.

Method:

A meeting was called of International Forestry Engineers and Irish Engineers with a knowledge of and experience in machinery development in the peat industry, forestry and agriculture. This team met for a three-day seminar in June 1980 which also involved a site and plantation inspection.

The general conclusions were that the most suitable particle size should be logs or billets not more than 15 cm long and 10 cm dia. with as low a moisture content as possible. It was generally agreed that the moisture content would be as high as 60% at harvesting but could be reduced to 35% with solar drying.

It was also agreed that harvesting should take place during the winter season, i.e. October to March when the moisture content of the material would be lowest and the subsequent summer season would be the most ideal drying period.

Having reached a decision on the final product and harvesting period various options on the harvesting system were discussed which can be summarised as follows:

1. Fell, chip, dry, transport and burn.
2. Fell, chip, transport, dry and burn.
3. Fell, dry, chip, transport and burn.
4. Fell, dry, bale, transport and burn.
5. Fell, bundle, dry, chip, transport and burn.

It was agreed that No. 1 option was the most desirable. Following a further discussion the following design criteria for the machine were agreed:

1. The prototype machine to be semi mounted on a standard tractor. The wheel gear or tracks to be capable of operating on a ground pressure of 0.105 kg/cm^2.
2. The machine to have a pay load capacity of 2 tonnes of material.
3. The machine to be capable of cutting twin rows of trees at 1 metre spacing between the rows and within the rows and should be designed so as to be capable of travelling in the three metre space provided between the twin rows.

4. The trees to be cleanly cut as close to the ground as is practical by a pair of power driven circular saws.
5. The machine should collect and convey the cut material into a chipper which would provide billets or particles not longer than 12 cm with a maximum diameter of 10 cm. This chipping mechanism would be a shearing device of slow rotating drum type consisting of a number of cutters positioned longitudinally on one roller and shearing against another counter rotating drum roller.
6. The cut or semi cut material to be deposited onto an agitator device designed to break off any particles that are not fully severed.
7. The cut material to be conveyed into a hopper located or semi-mounted on the vehicle.
8. The hopper to be either hydraulically tipped or have a chain unloader.
9. The transport size of the machine to be in line with E.E.C. vehicle size for transport regulations.
10. The machine to have a working speed of approximately 3 kilometres per hour.
11. The horsepower requirements of the machine to be calculated at a later date when the power requirements of the processing mechanism would be determined.

Design:

The above specifications were put on design and following further meetings and changes in design as problems were highlighted the final design was agreed in November 1980 as follows:

The harvester will comprise of a wheel frame with means at the front end for connection to tractor, a harvesting apparatus further comprising (all supported by the frame) a push bar disposed at a substantial height above ground level and laterally offset from the tractor connection for bending forward the trees for harvesting. A pair of horizontal cutting saws each having a diameter of 106.6 cm are located side by side close to the ground and to the rear of the push bar for cutting the trunk of each tree near ground level. A billeting mechanism located to the rear of the cutting saws and a means of conveying cut trees rearwardly from the cutting saws and feeding them into the billeting mechanism. There is also a further means of conveying the billets from the billeting mechanism to the hopper.

The harvesting apparatus frame further supports a flailing device to the rear of the billeting mechanism. The billets being passed into the flailing device for separation, trimming and splittling prior to conveyance to the hopper. The wheels of the frame will comprise a respective pair of bogie mounted tracked wheels on either side of the frame and all the apparatus components will be conventionally powered from the power take off of the tractor.

Design Details:

The harvesting apparatus comprises a main frame which supports all the operational components of the harvester to be described. The rear of the frame comprises on each side a respective pair of wheels, each pair of wheels having a track and being pivotted on the frame in the manner of a bogie mounting. The front end of the frame has a means for connecting to a tractor.

A subframe is pivotically mounted about a horizontal axis to the frame and rigidly supports a horizontal pushbar and a substantial height above ground level. The subframe and pushbar are laterally offset from the connection means so as to provide sufficient clearance to simultaneously harvest two rows of trees.

The function of the pushbar is to engage and bend forwardly the trees as the harvester is pulled forward thereby to present at the base of the trunk of each tree to one or other of the pair of generally horizontal saws disposed side by side close to the ground rearwardly of the pushover bar. Each saw is mounted at the bottom of the subframe for high speed rotation about a generally vertical axis by means of a hydraulic motor.

Immediately above the saws the subframe supports a pair of horizontal collectors disposed side in the form of endless bands having collecting fingers, each band passing around a pair of front and rear guide wheels respectively. The subframe further supports a pair of vertical feed rollers having longitudinal serrations, each feed roller being disposed coaxially above the rear guide wheel of a horizontal collector. The horizontal collector and the vertical feed rollers are driven by a further hydraulic motor via a chain drive. The chain drive drives the top end of the rollers which in turn drive the rear guide wheels (which also serve as drive wheels) of the horizontal collectors. The mountings for the rollers are spring loaded towards one another in conventional manner for accommodating variations in the diameter of the tree trunks.

To the rear of the conveying is the billeting mechanism which comprises a pair of cutter assemblies, mounted for rotation about parallel horizontal axes.

Specifically, each cutter assembly comprises an elongate support bar of square cross-section mounted for rotation about the respective axis in conventional bearings at each end of the bar. A plurality of outwardly projecting cutting blades (carbon steel) in the form of rectangular plates are rigidly bolted each to a respective side of the bar by means of bolts, each blade being reinforced by a respective backing plate interposed between the heads of the bolts and the blade. Each blade extends 91 cm along the full length of the bar, and several pairs of bolts distributed along the length of the bar are used to secure each blade. The blades of each assembly are thus tangential to a common notional cylinder coaxial with the axis and having a diameter equal to the width of the support bar. The cutting edge of each blade is parallel to the rotational axis of the relevant assembly.

The cutter assemblies are driven in synchronism in opposite directions of rotation (100 R.P.M.) by respective sprocket wheels secured to the ends of the bars. The sprocket wheels are themselves driven by respective sprocket chains from drive sprocket wheels respectively. The latter are coaxially secured to respective intermeshing gear wheels which are in turn driven by a drive gear wheel by a hydraulic motor.

The arrangement is such that during rotation as described

above successive pairs of blades, one from each cutter assembly, are simultaneously brought into the gap between the assemblies, the blades of each such pair approaching one another in the gap at a decreasing angle 0 of less than 180°. At their nearest approach the cutting edges of each pair of blades are closely adjacent one another, i.e. 6 mm gap.

In order to provide accurate positioning of the blades relative to the bar so as to precisely adjust the small gap between the blades at their closest approach, the apertures in each blade through which the bolts pass are designed to permit a limited movement of the blade in the tangent direction when all the bolts are slackened. Precise adjustment may then be carried out by screwing in or out distance-setting bolts whose heads bear against the inside surface of the adjacent blade. When the desired setting is achieved the bolts are locked in position by lock-nuts and the bolts tightened.

The cutting edge of each blade which is 7.6 cm deep is formed by a simple bevel on the trailing side of the outer edge of each plate.

Finally the subframe comprises a protective plate below the rear edges of the cutting discs.

The horizontal pivoting of the subframe on the main frame permits the angle of inclination of the subframe to the vertical and thus of the cutting saws to the horizontal to be adjusted from about 0-10°. This adjustment is achieved by a rigid coupling member of manually adjustable length connected between the subframe and the mainframe.

Construction:

Building of the prototype machine described above commenced in early January 1981. To date much of the conveying equipment, cutting mechanism and the pushbar system have been constructed. The billeting mechanism is almost complete and the machine will be ready for field testing by November 1st, 1981.

A 150 horsepower tractor will also be adapted (to reduce the bearing pressure to less than 0.282 kg/cm^2 for hauling the harvester.

DESIGN AND BUILDING OF A FORESTRY WASTES HARVESTER

Contract Number	: ESE-049-F
Duration	: 24 months
Total Budget	: 100 000 unités de compte CEC contribution 50 %
Head of Project	: LUCAS Jean, Ingénieur en chef du G.R.E.F. au CEMAGREF
Contractor	: Centre national du Machinisme agricole, du Génie rural, des Eaux et des Forêts
Address	: CEMAGREF Parc de Tourvoie 92160 ANTONY France Tél. : (1) 666 21 09

Summary :

Branches and other wastes let on the ground after the log harvest in French forests represent a potential of 4 TWh per year.

Collection of these wastes on a rentable scale for a grower means the design of a specific equipment.

Targets for the on-going design are :

- Flexibility (steering and slope potential)
- Wide scale of speed (100 meters per hour to 27 kilometers per hour)
- Resistance to shocks
- Non-easy collection of wastes potential
- Low costs compare to product selling prices

Calculations on steering, transmission, suspension, power generation, of the specific tractor and on sizing, grinding, conveying and storing of the product are under way.

. FRENCH FOREST AREA : $14,5 \times 10^6$ Ha.

. LOST FORESTRY WASTE EQUIVALENT TO 17 twh/year

. AFTER-FIRST-THINNING WASTE EQUIVALENT TO 4 twh/year

THIS PROJECT INTEND TO

RECOVER A SIGNIFICANT PART OF THIS LAST POTENTIAL FOR FUEL ON A COMMERCIAL SCALE.

THIS HARVEST OPERATION NEED A SPECIFIC EQUIPMENT

PROBLEM IDENTIFICATION

INPUT PRODUCT :

. IRREGULAR IN SHAPE AND SIZE BRANCHES

. BRANCHES GATHERED ON BUNCHES

. BUNCHES SCATTERED IN THE FOREST

. FOREST PARCELS SMALL AND SCATTERED

OUTPUT :

. CHIPS OR PELLETS OR PIECES OF BRANCHES

. STORED IN THE FOREST ALONG THE ROAD, OR IN THE FARM

. USED ON THE FARM OR SOLD.

DETAILED CONSTRAINTS AND CONSEQUENCES ON DESIGN

CONSTRAINTS :		CONSEQUENCES :
BRANCHES DIAMETER 1cm-15cm VERY INTECRATED BUNCHES OF BRANCHES	-------->	° NEED FOR A MORE UNIFORME AND EASY HANDLED PRODUCT ° VERY VARIABLE POWER REQUIREMENTS ° NO TRACTION ON BRANCHES BUT CUTTING IN THE BUNCH. ° FLEXIBILITY FOR ACCESS OF CUTTING TOOLS TO THE BRANCHES IN THE BUNCH
SCATTERED BUNCHES AMONG PROTECTED TRECS *(BECAUSE OF SEEDS AND WOOD)*	-------->	° TRANSPORTING CAPACITY ON THE VEHICULE ° EASY DISPLACEMENTS ON VARIOUSLY SLOPPED AND ORIENTED WAYS WHEN LOADED ° REDUCED WEIGHT HEIGHT AND WIDTH ° IMPORTANT VARIATION OF WORKING SPEED ° PROTECTION AGAINST VIOLENT HAZARDS
STRUCTURE OF FRENCH FOREST *(SMALL AND SCATTERED AREAS)*	-------->	° POTENTIAL FOR RUNNING HIGH SPEED ON ROADS

COMMERCIAL EXISTING SYSTEMS DO NOT MATCH THESE CONSTRAINTS

.... THEY ARE DESIGNED MAINLY TO CHIP OR CUT HARVESTED WOOD

.... THEY LACK OF POTENTIAL FOR

- THIS SPECIFIC CUTTING OPERATION
- THE COMPLEX DISPLACEMENT AT WORK IN PROTECTED AND SENSIBLE FOREST

.... THERE ARE PROBLEMS OF SIZE,
WEIGHT,
POWER REQUIREMENTS,
AND SPEEDS POTENTIAL

MAIN CONCEPTS

FOR THE DESIGN OF A SPECIFIC EQUIPMENT

TRACTOR :

- ° MAXIMUM HEIGHT = 2.6 M
- ° MAXIMUM WIDTH = 1.8 M
- ° 4 WHEELS INDIVIDUALLY STEERED AND POWERED.
- ° TIRES : INSTEAD OF TRACKS
- ° AUTOMATIC FRAME LEVELING
- ° 80 HORSEPOWER ENGINE
- ° 6 TONS WEIGHT
- ° CONTINUOUS VARIATION OF SPEED FROM + TO - 27 KM/H. (WITH NO VARIATION OF TRANSMISSION EFFICIENCY)
- ° MULTISENSE DRIVING FOR OPERATOR
- ° VERY LIGHT FRAME DESIGN STILL ABLE TO SUSTAIN FOREST HAZARDS.

LOADS :

6 M3 CONTAINER ON THE TRACTOR

CUTTING AND CHIPPING :

..... LIGHT, EASY HANDLED HEAD ABLE TO CUT AND CHIP IN THE BUNCH OF BRANCHES....

..... STUDY UNDERGOING IN ORDER TO

- . HIGHLY INCREASE CHIPPING EFFICIENCY
- . USE LOW WEIGH DEVICE ON THE CUTTING-CHIPPING HEAD
- . FEED AUTOMATICALLY THE PRODUCT INTO IT.

PREDICTED WORK ORGANIZATION

HARVEST 6 M3 OF CHIPPED BRANCHES

↓ *EVERY 1/2 HOUR*

EMPTY THEM INTO CLOSE, FIXED 20 M3 CONTAINERS

↓ *MIDDLE AND END OF THE DAY*

TAKE THEM TO THE FARM

EVERY DAY 5 hours of effective chipping during 8 hours of work.

ECONOMICS :

- 30 tons/day of wood or 15 tons/day dry matter
- with a price of the machine of 500 000 Francs.

this figures mean 530 Francs per ton of oil equivalent.

SUBJECT III. ALGAE

Methane production by mariculture on land

Investigations on the mass culture of marine algae in Southern Italy

Exploitation of lagoon macro-algae for biogas production

For a solar biotechnology based on microalgae

Biomass from offshore sea areas

Biomass from offshore sea areas

Renewable hydrocarbon production by cultivation of the green alga Botryococcus Braunii - Investigation of the factors affecting hydrocarbon production

METHANE PRODUCTION BY MARICULTURE ON LAND

Contract Numbers	: (1) ESE - R - 020 - D (until Feb. 28, 1981) (2) ESE - 460 (March 1, 1981 - ...)
Duration	: 12 months (each)
Total Budget	: DM 302,200 (1); DM 350.100 (2)
Head of Project	: Prof. Dr. K. Wagener
Contractor	: Technical University Aachen
Address	: Technical University Aachen Templergraben 55 5100 Aachen West Germany
Subcontractor	: CSMA - Instituto di Microbiologia Agraria e Technica dell'Universitá di Firenze

Summary

An analysis of mixing in mass cultures of microalgae showed that the requirement of mixing energy can go down to about 0.12 W per m^2 of pond area, or 4% of the energy stored in the harvest. Ponds for large-scale have been designed and tested as sectors of a round construction. Two ponds of 280 m^2 each are going to be installed at Lamezia. A system and cost analysis of various conversion routes for algae showed that the most economic concepts are systems with by-products credits, e.g. by selling the fermentation residue as high-value fertilizer, or part of the harvest as concentrated animal feed.

1 Goals of Energy Farming by Mariculture on Land (MCL)

- To avoid the pressure on land by using arid coastal areas and seawater
- To make use of the high productivity of phytoplancton
- To overcome the technological difficulties of open sea mariculture

2 Activities carried out during the last 9 months

(1) Biological work on algae growth: see report by Materassi et al.
(2) Work on low energy mixing procedures in algae ponds
(3) Large-scale pond design and construction
(4) Analyzing various conversion routes for algae and their economics

2.2 Summary of results on mixing in algae ponds

- Mixing intensity must be high enough to avoid settling and thermal stratification, and to induce the ability of proper settling rate for easy harvesting.
- To increase the productivity by more intensive mixing (due to the intermittent light effect) is out of discussion for energy farming:
  10 light/dark cycles per second would double the yield [KOK].
- Conclusion from the foregoing: mixing intensity as low as possible, in order to save mixing energy
- Tested mixing procedures: paddle wheel, waves, mixing boards (see Figure 1)

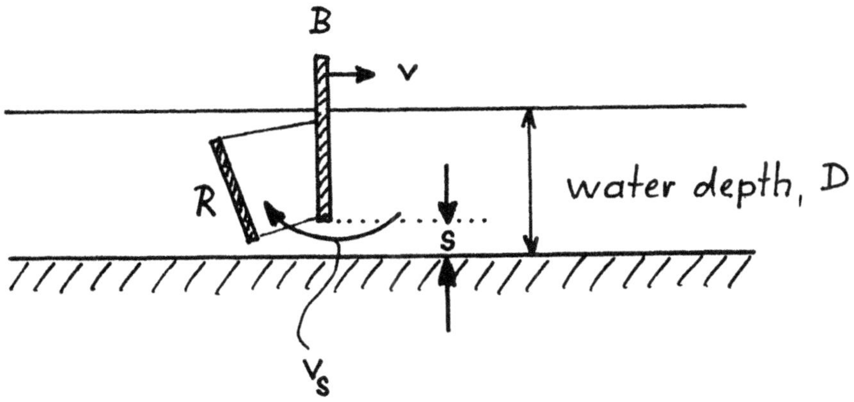

Figure 1: Mixing board. B= board, slowly moved with the speed v (∿6 cm/s); R = reflector, turns the water flow upwards; v_S = flow rate through the slit S = D·v/s.

- Comparison of mixing energy demand (30 cm water depth):

	W/m^2	% of fixed energy*
paddle wheel (flow rate = 15 cm/s)	4.6	143
artificial waves (one every 2 min)	0.25	8
mixing board (velocity 6 cm/s)	0.1	4

* 90 t/ha/yr; 18.8 MJ/t; conv. efficiency (thermal mechanical) = 0.3

2.3 Large-scale pond design: adapted to the easiest way of board mixing realization. Lamezia type: 2 X 280 m^2 each:

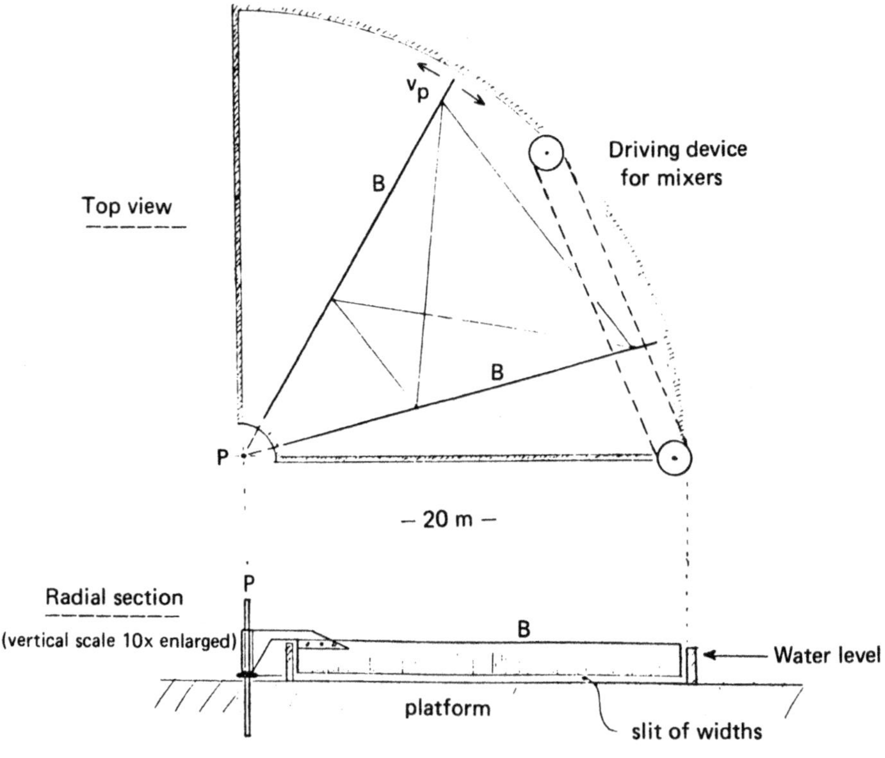

Figure 2: Mixing is accomplished by 2 boards, B, operating like windshield wipers of a car. The boards are moveable around the central pole, P, which is outside of the pond. They are interconnected to each other. Movement at the peripherical speed v_p forces the water to pass through the bottom slit of width S.

For R = 20 m (~ 280 m^2 of pond area), water depth D = 0.3 m, slit width S = 0.04 m, and v_p = 0.08 m/s, the tangential force to move both boards together is 13 kg (= 128 N) at the peripherical end of one board. The power input is 20 W (= twice the calculated amount, counting for friction and motor efficiency). This corresponds to 0.07 W/m^2.

Driving the mixers can be accomplished by means of an endless cable going around like a skilift, to which the mixer system is attached by another flexible cable or chain.

Radial mixing in the pond is efficient due to radial water flow, especially along the two radial walls, as has been tested experimentally.

2.4 Various conversion routes for algae and their economics

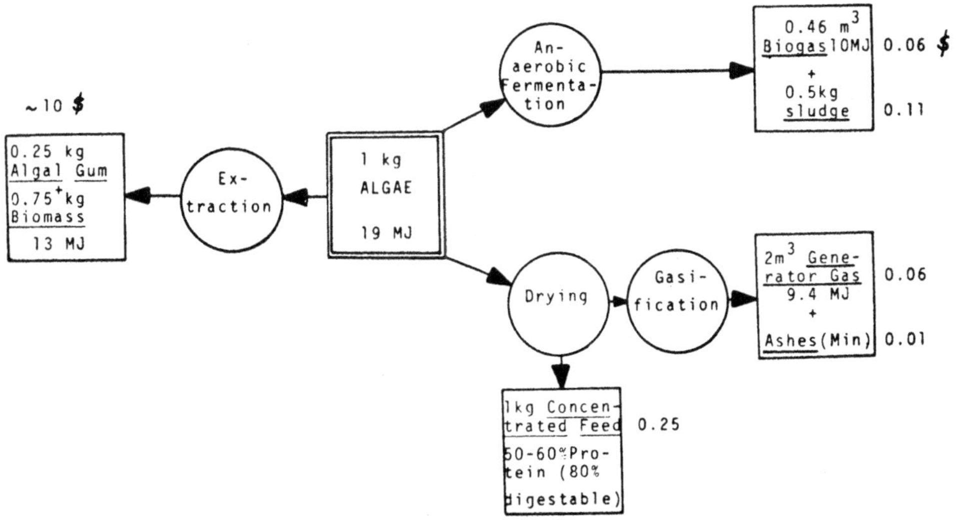

Figure 3: Various conversion routes for algae. Given are products, energy contents, other commercial uses and present market prices.

Conclusions - Perspectives

(1) Capital investment per m^2 of productive area (pond area) is comparable with agricultural land prices: ~5 $/$m^2$.
(2) In spite of their high productivity (90 t/ha/yr), algae are a too valuable product to be used just for energy.
(3) In terms of economics, the best concepts are combined systems producing biogas and sludge/fertilizer or biogas and part of the harvest as concentrated animal feed.
(4) In case external energy for seawater pumping is available (wind, ocean waves, solar; electricity), it is more economic to run an open system saving the valuable harvest as transportable energy carrier.

Table 1: Comparison of Energy Costs (June 1981)

(1 Dollar US = 2.30 DM)
(10 % interest p.a. on capital invest.)

Energy carrier	Source/Process/Conditions	$/GJ	Reference
Natural Gas	Consumer price	6	
Methane	From lignite (Lurgi process)	6.1	[1]
Methane	From hardcoal (Lurgi process)	8.7	[1]
Synthesis Gas	From lignite (Lurgi process)	6.7	[1]
Synthesis Gas	From hardcoal (Lurgi process)	9.6	[1]
MCl Biogas	Energy farm - closed system	17	
MCl Biogas	Energy farm - open system	13.4	
MCl Biogas	Energy/Fertilizer farm - closed syst.	14.8	
MCl Biogas	Energy/Fertilzer farm - open syst.	11.3	
Crude oil	OPEC	6.4	
Gasoline	Refinery gate, without tax (Germany)	9.1	
Methanol	From lignite (Winkler process)	7.4	[1]
Methanol	From hardcoal (Koppes-Dotzeck proc.)	11.7	[1]
Ethanol	From sugar cane (Brazil); without tax	17.7 (23)*	[2]
Methanol	From MCl methane - open system	20.4	
Electricity	Industrial consumer price	~24	

* in brackets: based on the actual ROI of 21 % p.a.

References

[1] Study prepared for the FRG Ministry of Research and Technology by DFVLR: Fossile Energieträger. Report E 1-3 (March 1979).

[2] Centro de Technologia Promon, Rio de Janeiro/Brazil: CTP Newsletter, vol. 5, nº3 (Aug. 1980).

"INVESTIGATIONS ON THE MASS CULTURE OF MARINE ALGAE IN SOUTHERN ITALY".

Authors : De Zarlo S., Tredici M., Balloni W. e Materassi R.

Contract Number : ESE-R-020-D

Duration : 12 months 1 March 1981 - 28 Feb. 1982

Total Budget : DM 354.000 CEC Contribution : 177.050 DM

Head of Project : Prof. K. Wagener

Subcontractor : Centro di Studio dei Microrganismi Autotrofi del CNR

Director : Prof. Gino Florenzano

Address : Piazzale delle Cascine, 27 - 50144 FIRENZE Italia

Summary

This paper reports the results of the research on the mass culture of phytoplanton organisms in fertilized seawater carried out in the south of Italy (Lamezia Terme - Calabria) during 1980 and the first five months of 1981. The main goals of this work have been the evaluation of the productivity of marine photosynthetic organisms using Tetraselmis sp. as a model organism, the development of the appropriate culture technique and the selection of algal strains with growth properties suitable for the MCL process. Preliminary results on the mass culture of two "filtrable" strains are reported.

1.1 Introduction

The research activity of the "Centro di Studio dei Microrganismi Autotrofi del C.N.R. di Firenze" during 1980 and the first half of 1981 has been devoted to the evaluation of the productivity of marine photosynthetic organisms grown in fertilized seawater in outdoor mass culture under the environmental conditions of southern Italy. During this period further experiments were made in order to test the influence exerted by some relevant parameters, as the dilution rate of the culture and the concentration of nitrogen and phosphorous, on productivity.

Another important point under investigation concerns the selection of algal strains with growth properties suitable for the MCL process. In 1980 several organisms were isolated from seawater and their growth tested under laboratory conditions and in small outdoor culture. Two of these organisms have been selected for further investigation during 1981.

1.2 Materials and methods

The experimental station at Lamezia Terme (Calabria) has been described in previous reports (1-2). Outdoor culture facilities have been enlarged with the installation of two small circular ponds (1 sq metre of illuminated surface each) and four ponds having 2 sq metres of illuminated area each. In order to test the possibility of mixing the culture at the expense of the aeolian energy, the paddle wheel of these ponds are moved by a windmill.

Four 30 litres digesters have been built; they will be utilized to test the biomethanation of the biomass from different algal species.

1.3 Results

1.3.1 Productivity data

The data concerning the productivity of outdoor mass cultures of *Tetraselmis* for the year 1980 are summarized in table 1. Each figure represents the mean productivity of at least two ponds. The mean value of the yield for the whole year has been 17,6 g of dry weight/sq meter/day, corresponding to the production of more than 64 tons of biomass/ha/year.

During 1980 it has been confirmed that the highest productivity of *Tetraselmis* cultures are reached when the concentration of biomass is maintained within 55-70 g of dry matter/sq metre in summer. When the amount of solar radiation falls below 500 cal/cm^2/day the optimum biomass concentration in the ponds is a little less. In order to keep the concentration of biomass within the above mentioned values, in summer it is necessary to increase the dilution rate up to 0,5 per day.

During 1980 the carbon requirement of the cultures was

Table 1 - Yields of outdoor mass cultures of *Tetraselmis* sp. carried out in Lamezia Terme during 1980.

month	mean air temperature min.	max (C)	total solar radiation $cal/cm^2/day$	productivity (g dry weight/ m^2/day)
January	5,2	12,8	176	5
February	4,6	13,8	254	7
March	6,5	14,6	338	11
April	6,5	15,3	452	16
May	11,5	18,5	565	25
June	14,1	24,3	620	29
July	16,3	26,7	628	36
August	18,6	28,7	566	30
September	14,1	26,3	448	23
October	11,1	21,9	313	16
November	10,2	18,4	216	8
December	5,9	13,6	153	6

neet by bubbling pure CO_2 from cylinders. In this way precipitation of inorganic material was prevented.

Since during 1980 the seawater was enriched only with inorganic nutrients, the yields obtained represent the true photoautotrophic productivity of the *Tetraselmis* cultures in southern Italy.

Two years and half of experiments with outdoor cultures of *Tetraselmis* have demonstrated that this species has a remarkable stability, being able to withstand unfavourable climatic conditions. Contamination with other algae are relatively uncommon, expecially with high dilution rate and under the vigorous stirring of the paddle wheel. Frequent contamination occurs in April and May by several pennate diatoms. These organisms grow mostly on the surface of the paddle and on the wall of the ponds at the air-liquid interface.

1.3.2 Selection and testing of different algal species

Although *Tetraselmis* has many favourable properties for outdoor mass culture operations, it is difficult to harvest, because it is not "filtrable" (it cannot be separated with a strainer) and the sedimentation rate is not fast enough unless some chemical flocculant is added to the culture.

In view of these difficulties, a search for marine algae that can be separated efficiently and economically from the culture liquid has been undertaken. Several filamentous or colonial strains have been isolated from the seawater of the S.Eufemia Gulf. Two of these organisms have been selected for outdoor culture experiments. One of these is a blue-green bacterium of the genus Oscillatoria, close to O. raciborschii (figure 1), forming very long trichomes (more tham 2 mmetres long). This organism can be easily separated from the seawater whem the culture is poured on a nylon net (figure 2) with a 60-80 micrometres mesh. The slurry obtained reach a concentration of total solid of 7-8% withoout further treatments. Oscillatoria cells possess gas vacuoles that lower the sedimentation rate of the organism, so that the settling of the biomass can be prevented with a moderate mixing intensity. Oscillatoria grow best at pH values above 8,5; the growth on fertilized seawater is comparable to the growth observed on the artificial culture media suitable for Spirulina and containing high amount of carbonate-bicarbonate. Oscillatoria is being successfully grown in outdoor mass culture; the yields are about 30% lower compared with the yields of Tetraselmis. We are of the opinion that the optimal conditions for the mass culture of this species have not been found up to this moment.

The other organism now under investigation in outdoor culture is a green alga classified as Ulva sp. This species forms a macroscopic thallus, provided with special structures for attachment to the substrate. In stirred cultures Ulva grows in suspension. In this situation the settling rate during the day is low, probably by virtue of small oxigen bubbles entrapped in the thallus.

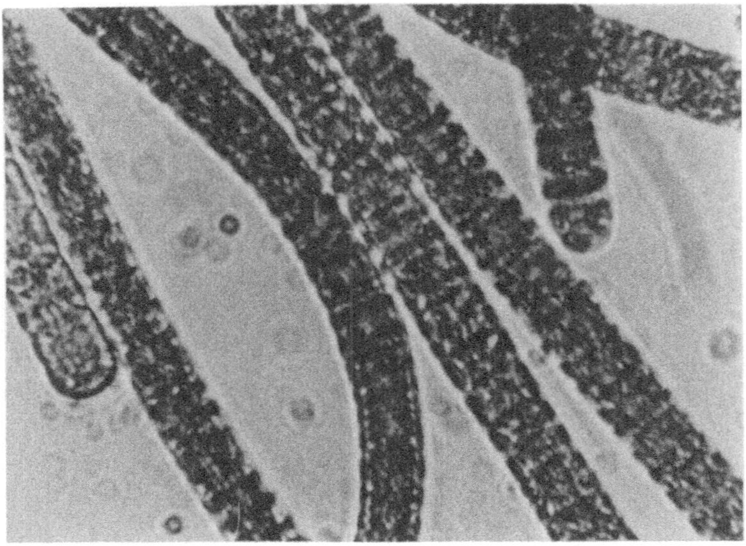

Figure 1 - Micromorphology of Oscillatoria raciborschii (x 800)

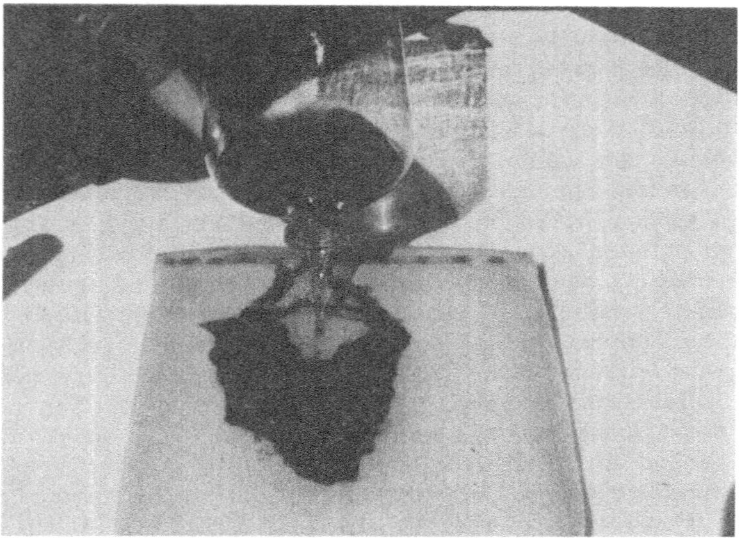

Figure 2 - Harvesting of O. raciborschii by fil tration on a nylon net.

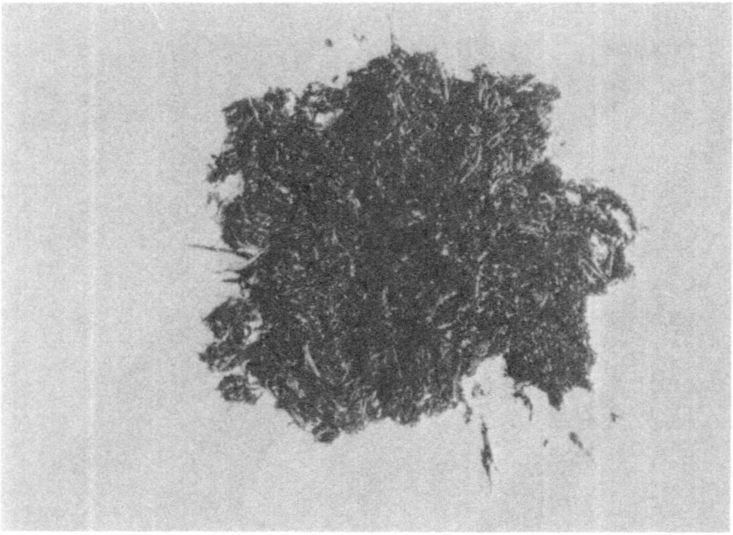

Figure 3 - Biomass of Ulva sp. harvested with a wi re net.

The separation of Ulva biomass from the culture medium can be achieved very easily with a wire net immersed in the culture. In April the productivity of Ulva sp. was comparable with that of Tetraselmis. It has been shown that with a low retention time of the sewater in the ponds, Ulva cultures can be preserved from contamination by unicellular algae. This is a good property, since in the MCL process very high dilution rates are adopted. Ulva is a typical marine organism, because it grows well only on fertilized seawater.

After separation from the culture solution the excess water is quicky removed from the biomass (Figure 3). This can be spread on a net and sun dried in few hours.

2. Conclusions

The investigations on the outdoor mass culture of Tetraselmis sp. have demonstrated that, under the climatic conditions of southern Italy, high productivity of biomass can be obtained from marine algae grown on fertilized seawater. At Lamezia Terme yields of about 65 tons of dry matter/hectare/year have been obtained. From these results it is possible to estimate that under more favourable climatic conditions (butter distribution of the solar radiation and absence of a cool season) it is possible to attain 90-100 tons of dry biomass/hectare/year.

It has also been demonstrated that algal cultures in seawater can be maintained for more than one year without apreciable loss of productivity, applying the appropriate culture technique. Hence the basic biological assumptions of the MCL process have been verified successfully. The work in progress is directed to the selection of algal species with good properties: first of all chemical composition of biomass and easy harvesting. The culture of "filtrable" organisms offers interesting possibility for appliing measures directed to an effective control of the species composition.

3. References

(1) Florenzano G., Materassi R., Balloni W., Mozzoni M. (1978) Investigation in 1978 on the productivity of marine algae in seawater for the project "Methane production by mariculture on land". 2 nd Coordination Meeting of Contractors, EC Solar Energy Programme, Project E. Brusselle, 9-10 November

(2) Florenzano G., Materassi R., Balloni W., Mozzoni M. (1979) Investigations on the productivity of outdoor mass cultures of marine algae in fertilized seawater for the project "Methane production by mariculture on land". 3 rd Coordination Meeting of Contractors, EC Solar Energy Programme, Project E. Taormina 6-7 June.

EXPLOITATION OF LAGOON MACRO-ALGAE FOR BIOGAS PRODUCTION

Contract number : ESE-R-039-I

Duration : from July 1980 to June 1983

Project leader : Prof. U. CROATTO

Contractor : C.S.A.R.E. - Centro per gli Studi e le Applicazioni delle Risorse Energetiche

Address : Corso del Popolo 261 - 45100 ROVIGO (Italy)

Summary

Environmental factors, such as concentration of nutrigents, oxygen content, temperature and turbulence of waters, influence the growth of micro- and macro-algae in lagoons.

In a previous study, we have considered the macro-algae productivity and the location of natural accumulation areas in the Venice lagoon (500 km^2).

Two main species of macroalgae (*Gracillaria Confervoides* and *Ulva Rigida*) are lying on the lagoon bottom and accumulate in correspondence to the subsidences, due to tide movements.

The total macroalgae availability, referred to winter time (minimum productivity), is about 1.0 kg/m^2 : the estimated value for biomass reaches about 500 000 tons (weight of wet dripped matter). In the warmer seasons the total amount reaches a value of 2.5 millions of tons.

Algae can be removed from the accumulation areas without producting any ecological unbalance. It has been found that accumulation areas have low depth; macroalgae are note anchored to the bottom of the lagoon. Simple collecting and transportation devices have been devised. Wet algae can be stored for a short period of time in delimited areas of the lagoon to assure their living.

We are now pursuing researches about mapping, harvesting, storage and pretreatment of lagoon algae and their transformation to biogas. This requires the determination, through laboratory and pilot plant experiments, of the fundamental parameters for industrial plant design and production.

PRELIMINARY REPORT

Some data are initially given which allow us to evaluate the mean annual production of algae in the Venice lagoon. The results in the first stage of the research are then reported : these concern laboratory experiments, anaerobic fermentation carried out in a pilot reactor, analytical determination on the biogas and the fermentation residues.

Mesurement equipment

MAPPING : A prototype of shalloon draught motor-boat has been used to exploit the different areas of the lagoon.

ANAEROBIC BIOGAS PRODUCTION : A pilot anaerobic reactor, 500 l of capacity, vapor heated, equipped with a low speed stirrer, connected with a gasometer, 5 000 l of capacity, has been used.

ANALYSES : Samples of biogas have been tested by gas chromatographic techniques (Hewlett Packard Mod. 5710/A linked to a Mod. 3385/A microcomputer).

The metal content in solid residues has been determined by atomic absorption techniques (Perkin Elmer Mod. 503).

Macroalgae availability in the Venice lagoon

A mapping of the Venice lagoon is in progress. The results obtained so far allow an estimate of the natural abundance of macroalgae (*Ulva* and *Gracillaria*). The resident overall quantity presents a seasonal variation between 0.5 and 2.5 millions of tons (wet algae). Maximum growth occurs during spring time, the productivity starts rising in February to reach its highest point between April and May, while September it starts going down again. Also the quantitative ratio between the two species of algae changes with seasons.

On taking into account the growth rate of these algae in the lagoon environment, we have estimated an annual availability of about one million of tons of wet algae without producing any appreciable unbalance in the ecological system. This corresponds to a mean daily collection of ca. 2 700 tons of wet algae which can be carried out with about ten vessels suitable for operating in low depths.

Harvesting and storing techniques

A prototype of shalloon draught boat, equipped with a mechanical device for harvesting the algae from the lagoon, is in an advanced stage of planning. We forecast that the experiments in the lagoon will be carried out within 1982. These experiments must supply the basic data for the final design of the harvesting vessels.

The storing problems are essentially connected with the difficulties of algae collection during the storming weather. Consequently, a storage for about one week is foreseen in a well delimited area of the lagoon, where optimal living conditions for algae should be ensured. This area should be chosen close to the shore and the plant production. Longer storage times will be chosen according to the result of the pilot plant experiments.

Biogas production

Laboratory tests, with mesophylic bacteria (operating at ca. 25-30 °C), have been carried out in a pilot reactor with the aim to find the optimal operating conditions as regards inoculation problems and temperature and stirring conditions. We have considered the mesophylic bacteria instead of thermophylic ones, since these latter during winter time of Northern Italy should require a very high fuel consumption. With the mesophylic bacteria the reactor will be heated in winter time consuming about 20 % of the biogas produced. The use of lyophilized bacteria addition into the reactor has not given any appreciable increase in the biogas production. On the other hand the saline content did not show any remarkable influence on the fermentation process.

As far as the analytical aspects are concerned, gas chromatographic methods have been set-up for the qualitative and quantitative determination of the chemical substances obtained (volatile inorganic substances, oxygenated organic derivatives, hydrocarbons).

The preliminary results obtained allow us to estimate a yield of biogas production (with about 70 % in volume of methane) of 0.35 m^3/kg of dry algae.

For the production plant, using one million of tons of wet algae, about ten reactors of 2 500 m^3 of volume should be forecasted.

We have also studied the possible removal of CO_2 and small amounts of H_2S by countercurrent treatment of biogas with monoethanolamine solution.

The influence of the size of the algae particles and of the stirring of the reacting biomass on the rate of biogas production has been studied. The optimal condition for the biogas production is represented by a constant temperature, composition and ground size of algae in the bulk of the reacting biomass. Consequently, one of the objectives of the present laboratory experiments is to define the pretreatments to be carried out on the algaal biomass before introducing it into the anaerobic reactor, taking into account the storage conditions of the algae.

On the basis of the experimental results, we can estimate that the fermentive utilization of the algae should be of about 50 %, since the fermentation rate becomes progressively lower.

Solid residues from anaerobic digestion

The solid residues of the fermentation consist of about 50 % of the unused algae and of the residual biological sludge. The fate of these residues, depending on the degree of the pollutants contained (heavy metals), might be used in agricultural lands, or a controlled disposal or incineration.

Experiments have been carried out by feeding the anaerobic reactor with algae harvested in different locations of the Venetian lagoon (facing the Industrial area, the fresh water canals interested by an agricultural area of 200 000 hectars, and the historical center of Venice).

As far as the analytical aspects are concerned, tests conducted by Atomic Absorbtion Spectrometry on the fermentation residues, gave qualitative and quantitative values remarkably different as concerns the heavy metal content; a fairly good correspondence has been found between the metal concentration we have measured and the findings on the sediments of the lagoon as reported by other authors. This results indicate that macroalgae can be considered reliable bio-indicators.

Researches are now in progress to determine the nutrient capacity on the residues in the fertilization of lands cultivated with Indian corn, with their pullutants content being of course taken into account.

Final comments and conclusions

The researches carried out so far clearly indicate the possibility and the opportunity of going on to the realization of a production plant, which treats one million of tons of algae as a year average.

The problems which must be very carefully examined are : the most suitable locations for the plant installation and for algae storage; the realization of the collecting vessels and the algae availability during the winter period, which could cause fluctuations in the monthly production of biogas.

The biological utilization of the lagoon macroalgae will also contribute in reducing the lagoon pollutants (nitrogen, phophorus, heavy metals, etc.) and the algaal blooming phenomena, which occur twice a year and cause a decrease in the oxygen content, algaal decomposition and fish mortality. Consequently, evaluating the operating costs, which have become already acceptable in view of the steadily increasing fuel costs, one must also consider the fringe benefits of environmental character.

FOR A SOLAR BIOTECHNOLOGY BASED ON MICROALGAE

Authors : C. GUDIN, D. CHAUMONT

Contract Number : ESE-R-037-F

Duration : From June 1978 to June 1980
From June 1980 to June 1981

Total Budget :	1978-79	1.151.776	CEC contribution	230.700
	1979-80	543.927		117.000
	1980-81	470.000		234.000
		2.164.693		581.700

Head of Project : C. GUDIN

Contractor : Société Française des Pétroles BP

Address : SFBP, 10, quai Paul Doumer - 92401 - COURBEVOIE Cedex France

Summary

The initial target was to establish the maximum biomass production per unit of time and surface in combining photosynthetic bacteria and microalgae in double layer systems of cultivation. Having shown this technological possibility we have concentrated our effort on setting up a fully automatic system of microalgae monoculture with record and control of all parameters :

1) Yearly average obtained for a complete year is 20.9 g/m2/day corresponding to a photoconversion yield of 4 %.

2) In mediterranean sunlight conditions the theoretical maximum based on a 6.6. % photoconversion yield would be 34.5 g/m2/day as a yearly average.

3) Energetic evaluation of the production leads to a value $\leqslant 1$, showing no beneficial fixation of energy.

4) Economic evaluation of production cost indicates a cost of $\simeq$ 5 FF/kg of biomass.

Keeping in mind these 4 conclusions our recent targets have been :

1) To developp all technologies able to improve drastically the fixation of energy.

2) To select strains able to give a final product $\geqslant$ 10 FF/kg.

To reach the 1st target, we have started

1) a 3 steps experimental programme (June 81 → June 84) on 1 m2, 10 m2 and 100 m2 pilot plant tubular cultivator.

2) a new technological way based on "cells immobilized on a support" for production of an exocellular biomass.

In the frame work of the 2nd target we have selected 2 types of strains :

1) Red microalgae producing high value polysaccharides (exocellular biomass) and cellular biomass for methanisation (cooperation with ESE/021 B)

2) *Botryococcus braunii* (cooperation with ESE/003/F) producing exocellular hydrocarbons and cellular biomass for methanisation.

This project will go on from the 1st June of 81 in the new "Solar Biotechnological Unit" set up by C.E.A. in Cadarache (Biological Department Radio Agronomie Service) Saint-Paul-lez-Durance - 13115 - France.

Introduction

A comprehensive report of the work done between June 1978 to June 1980 has been published with CEE agreement. (1) (2) (3).

In those publications and precedent internal reports (4), equipment and measures have been described. We will concentrate in this paper on the different results obtained between June 80 to June 81.

The target was, as defined by CE "examination of the various ways to get a favourable energetic and economic balances for algae production in tubular devices, to fullfill some food, chemical and biofuel demands and to cooperate with Mariculture on Land Project (34/D + I) and *Botryococcus* Project (03/F).

I - COOPERATION TO MARICULTURE ON LAND PROJECT

1.1. Growth of *Enteromorpha lingulata* in laboratory conditions (Prof. Wagener strain).

At 25° C, with a sea water medium (Na NO_3 170 mg/l, $K_2HPO_4(3H_2O)$ 43.76mg/l at pH 7.5) an aeration of 10 V/V/H (volume of air per volume of culture per hour) enriched with 1 % of CO_2 and 30 W/m2 of visible artificial light ($\simeq$ 618 Kcal/m2/day) we have obtained a doubling time of dry biomass of 2.5 to 3 days. The maximum dry biomass concentration obtained in these conditions was $\simeq$ 1 g/l after 2 weeks of cultivation starting with an inoculum of $\simeq$ 0.1 g/l. This production leads to $\simeq$ 1.4 /m2/day of biomass. By extrapolation of laboratory conditions, knowing our mediterranean conditions we have derived the maximum value of 10 t/ha/year.

1.2. Growth of *Synechocystis sp* (Lamezia strain I. L1) in laboratory conditions

The culture medium for 1 l of sea water was made of 0.08 g of Urea, 0.03 g of Na glycerophosphate 0.5 mg of $MgCl_2$, $4H_2O$ and 3 mg of Fe EDTA at pH $\simeq$ 8.

Experimentation has been done in continuous cultivation in a small 1.2 l cultivator with continuous artificial light (30 W/m2 of visible light).

Synechocystis sp can grow well in the range of 25° C to 42° C and can be considered as a thermophilic strain. At 32° C the best biomass doubling time was $\leqslant$ 8 hours. For residential times of 24 h, 12 h, 8 h, the productivities were 15 g/m2/day, 13.8 g/m2/day and 10.6 g/m2/day.

Applying the same type of extrapolation than before we can derive 47 t/ha /year.

1.3. Growth of *Spirulina maxima* (IFP strain) in laboratory conditions

The culture medium has been described by Zarrouk (5). All experimentation has been done as previously in continuous cultivation (40 "steady states" of a minimum of a 1 week each).

We have studied t^0 (between 22° C to 42° C). Residential time (between 24 h to 8 h). Light intensity (8.9 W/m2 to 26.6 W/m2). NaCl concentration (11.5 g/l to 35 g/l).

Best results in biomass productivity and photoconversion yield are obtained at 32° C with a residential time of 21 hours. A quantum deman of 23 photons/mole of CO_2 fixed is inclosed agreement with Moyse results (6).

In our culture conditions the best biomass duplication time is 8.7 hours. We have firmly established that between (100 Kcal/m2 to 700 Kcal/m2 of light energy) →

the biomass concentration was linearly proportional : independently of light intensity (26 W/m2, 23 W/m2, 18 W/m2, 9 W/m2) or photoperiod lenght (24 h, 16 h, 12 h, 8 h).

Optimal temperature is 32° C and the strain is not thermophilic (not growing at 42° C and very slowly at 22° C).

Results obtained have proved it was possible by using continuous culture with an increased concentration of NaCl to adapte *Spirulina* to grow well on a medium containing 35 g/l of NaCl. A way of selecting strains adapted to sea water seems to be opened.

The best productivities obtained at 32° C on Zarrouk medium are $\simeq$ 10 g/m2/day in our conditions and definitely lower than for *Synechocystis*.

As far as we know for the moment *Synechocystis sp* is the most relevant strain for MCL in Lamezia being a sea water strain, very fast to grow, thermophilic with a good biomass productivity.

II - COOPERATION TO *Botryococcus braunii* PROJECT FOR HYDROCARBON PRODUCTION

We have done a great effort on the study in laboratory of the growth kinetic of *Botryococcus braunii* and on the influence of nitrogen and phosphorous nutrition. The results are available in the 03/F report.

III - COOPERATION ON METHANISATION OF THE CELLULAR BIOMASS PRODUCED IN THE TUBULAR CULTIVATOR (ESE/021 B)

Different samples of microalgae grown in on cultivator have been harvested by centrifugation an the fresh paste send to Prof. Naveau for methanisation study : *Scenedesmus acutus* and *Porphyridium cruentum*. The *P. Cruentum* cellular biomass was send after complete extraction of the high value exocellular polysaccharides. The CH_4 produced could improve the energy balance and the polysaccharide could solved the economic of the system. We are going to work now on this basis at the 10 m2 and 100 m2 level.

IV - DEVELOPMENT OF A BIOPHOTOREACTOR BASED ON *P. cruentum* CELLS IMMOBILIZED ON POLYURETHANE MOUSSE

P. cruentum in nitrogen starvation is able to produce an exocellular sulphated polysaccharide of high molecular weight (> 10^6) water soluble and acting as a thickening agent (7). We have entraped a liquid culture of *P. cruentum* in a polyurethane mousse and set up an illuminated biophotoreactor of 2 liters. After 14 months of continuous circulation of mineral sea water medium, the cells are still producing high concentration of polysaccharides ($\geqslant$ 4 g/l).

The production is followed by the measure of viscosity and the level of mousse colonisation and physiological state of the cells by optical and electronical microscopy.

The main difficulty is to obtain a combination of transparency and friability of the mousse to get a good filling of the column with a relevant surface/volume report.

A second photoreactor improved from this point of view has been made operational 3 weeks ago.

This exocellular production appears as a photo polymerisation of CO_2 in polysaccharide which is a way as a beginning in managing cell photosynthesis. Most of the light energy fixed is involved in the final (8) product easy to harvest by ethanol precipitation.

It is hoped, this concept of exocellular production will be applicable to *Botryococcus braunii* later on, for production of hydrocarbon.

This new technological concept solves in the same time the "energy balance problem" and the "economic aspect" of sunlight bioconversion.

We intend to do more basic research in this field to prepare the 2nd generation of light bioconvertors.

CONCLUSIONS

1) With a cost of production of $\simeq$ 5 FF/kg of biomass in our tubular cultivator using *Porphyridium cruentum* we are in a position to suggest an economic target for microalgae biomass (market price between 15 to 30 FF/kg of polysaccharide) in small size production unit ($\geqslant$ 1 hectare) compatible with the size of the market.

The 50 % of cellular biomass with low market value ($\leqslant$ 2 FF/kg as a food) can give methane and improved the energy balance.

2) Immobilized cells of microalgae producing exocellular biomass as *P. cruentum* has been successfully used for 14 months.

This solar biotechnology opens a new way to fullfill food, chemical or biofuel demands with photosynthetic cells.

3) Our laboratory has been able to help in solving some physiological and cultural problems on microalgae. *Synechocystis sp* growth characteristics have been established (MCL project) and improvment in growing *Botryococcus braunii* have been achieved.

(1) C. Gudin & col.-Rev. Internat. d'Heliotech. COMPLES- 2e semestre 1980, p. 55 to 62.

(2) C. Gudin & col.-Biochem. Soc. Transactions 8 (4) p. 481-482 (1980).

(3) C. Gudin & col.-Energy from biomass inter. conf. Brighton 1980. "new concepts in solar biotechnology".

(4) C. Gudin & col.-EC solar energy. Project E International report of ESE/007/F of the work from June 78 to June 80 (partly published in (1))

(5) C. Zarrouk -Contribution à l'étude de *Spirulina maxima* (Thèse de Doct. Sc. appl. Paris 1966).

(6) E. Guerin-Dumartrait & A. Moyse -Ann. Nutr. Al. 29 (16) 489-496 (1975)

(7) J. Ramus & D. Robins -J. Phyc. 1975 II p. 70-74.

(8) C. Gudin & D. Thomas -"Mise au point d'un biophotoreacteur à *P. cruentum* produisant des polysaccharides". C.R. Ac. Sc. (Juin 1981) (sous presse).

BIOMASS FROM OFFSHORE SEA AREAS

Contract Number	: ESE-R-021-UK*	
Duration	: 33 months	1 Oct 1980 - 30 June 1983
Total Budget	: £ 133,825	CEC Contribution : £ 60,000

Head of Project : Dr Joanna M. Jones, University of Liverpool
Researcher : T.J. Holt

Contractor : University of Liverpool

Address : Department of Marine Biology
University of Liverpool
Port Erin
Isle of Man

Summary

Ropes attached to concrete blocks and buoys have been placed in the sea in such a way that long horizontal protions are stable at pre-determined depths. Vertical portions exposed since February and March have been naturally 'seeded' by *Alaria esculenta* and *Laminaria saccharina*, with a peak in numbers between 6 and 15 m depth. The success of the former, cold water, species suggests it as a possible early crop. The horizontal portions were artificially 'seeded' with string bearing *Saccorhiza polyschides* gametophytes. This was not very successful partly because the string parted from the rope and partly because of poor development of the young sporophytes, probably due to low temperature. It seems that controlled conditions are necessary for successful artificial 'seeding'. Samples of *S. polyschides* have been collected and dried regularly for chemical analysis in Norway. The rate of drying is high at slightly elevated room temperatures.

*See also report from Nottingham University (joint contract).

1.1 Introduction

The aim of this project is to investigate the feasibility of growing kelp on artificial structures as an alternative energy source, since it is believed that if sufficiently cheap structures can be designed, and 2 kg dry weight of seaweed produced per square metre of sea surface, this would be both economically and energetically viable.

1.2 Techniques and Results

1.2.1 The Structures:

In this, the first year of the project, very simple rope structures have been placed in the sea, in 16 - 21 metres of water (tidal range). These consist each of a 50m horizontal rope suspended between a pair of vertical ropes each of which is attached to a concrete block below and a buoy on the surface. The horizontal rope is at 0, 1 or 2 m depth. The latter two are weighted with pieces of chain directly beneath supporting buoys. Small (0.064 m^3) concrete blocks have been used so far. A better anchoring system is being investigated - either by using more weight, e.g. disused car engine blocks, or by burying the concrete blocks in the substrate using an airlift system. Preliminary observations suggest the latter method may be suitable on deep sand but not on shingle. The strength of anchorage is being tested using a strain guage.

1.2.2 Seeding techniques

Whilst it is possible that merely placing the ropes in the sea and allowing them to 'self seed' could be adequate (as demonstrated by many mooring ropes which get covered in kelps in the space of 6 months), a seeding method is being sought which would allow selection of the algae grown. If successful this would ensure that growth occurred every year, but would entail culturing under optimal conditions before the young plants were put in the sea. Because of their high growth rates and relatively large size, *Laminaria saccharina*, *Saccorhiza polyschides* and *Alaria esculenta* have been selected.

The first method tried involved placing sporing plants over string covered frames in filtered running sea-water in outdoor tanks. The only contaminating plants found in one month of culturing were diatoms which were easily washed off with a hosepipe. This process had no effect on the developing gametophytes of *Saccorhiza*, which were observed in large numbers on the string although they developed very slowly due probably to the low temperature of the water (6 - 10°C). Unfortunately this method produced a very patchy distribution on the string and in future a spore suspension will be produced and poured onto the string.

In early April, after 4 weeks of culture, 25 metres of rope was 'seeded' at each of the 3 depths, (0, 1 and 2 metres) by coiling the string around the rope in situ and tying it at intervals. This method was not successful because the string did not remain in close contact with the rope.

Gametophytes are now being cultured under controlled conditions on small lengths (ca 30 cm) of string which will be tied individually to the rope at various intervals.

1.2.3 Growth

Plants are being monitored on both the vertical (unseeded) and horizontal (seeded) ropes.

1.2.3.1 Vertical ropes

These were placed in the sea on February 17th, February 25th, March 4th and April 1st.

All of the ropes produced a thick growth of diatoms within 2 weeks of being placed in the sea. This was mainly confined to the top 8 - 10 m. Green algae began to appear in small amounts with even smaller numbers of reds. The first laminariacean sporophytes were found on the February and March ropes in small numbers (about 6 per rope) and consisted mainly of Alaria esculenta and Laminaria saccharina. By May 7th some of them had reached 50 cm in length. Quite large numbers of plants of Desmarestia viridis had also appeared on the upper two-thirds of the rope.

The numbers of laminariacean sporophytes within 2 size classes on each metre of rope are being counted. Species identification is not being attempted on those plants under 10 cm. These counts are made in situ by divers and involve considerable inaccuracies.

No sporophytes have been found on the April ropes. This may reflect the cessation of sporing by the algae concerned. The numbers on three of the ropes put in the sea in February and March and counted on May 29th and June 9th are shown in Fig. 1. Only small numbers of plants appeared in May but numbers had increased by early June. Laminaria saccharina and Alaria esculenta have dominated this early period. The latter is favoured by winter temperatures (1) and few have appeared on the March rope although this may reflect a lack of spores in March.

There is generally a peak of production between 6 and 15 m especially in L. saccharina (Fig. 1). Possible explanations for the low numbers near the surface are i) the top 3 - 4 metres of rope needed to be hauled in over the side of the boat during seeding; ii) the diatom and green algal growth is thickest nearest the surface and may have led to reduced light levels in the early stages, as well as difficulty in spore settling; iii) high light levels before the diatom growth thickened may have inhibited the algae; iv) zoospores of subtidal species may be less likely to be released into surface waters than into water of more than the tidal range in depth.

The plants on these ropes will be monitored until October, providing growth data throughout the main growth season. The ropes will then be harvested and production in terms of wet, dry and organic weights calculated.

1.2.3.2 Horizontal ropes

The seeded strings (Fig. 2) have so far produced only a few sporophytes, although more than have been produced by natural seeding on the rope. The surface rope is the only one on which all of the seed string has been retained and portions of this have been tied closely to the rope to encourage the sporelings to attach themselves to it.

Again these ropes will be monitored until October. The seeded string in culture at present will be placed on the ropes in early July and followed in a similar manner.

1.2.4 Chemical constituents

While a considerable amount is known about the seasonal variation in chemical constituents of Laminaria saccharina and Alaria esculenta (2, 3) little is known about Saccorhiza polyschides. Dry weight and ash weight affect the amount of organic material available for energy and some organic constituents (e.g. mannitol and algin) are more suitable than others (e.g. cellulose and protein) (4). It is important to gain information on this so that the species grown and times of harvesting can be

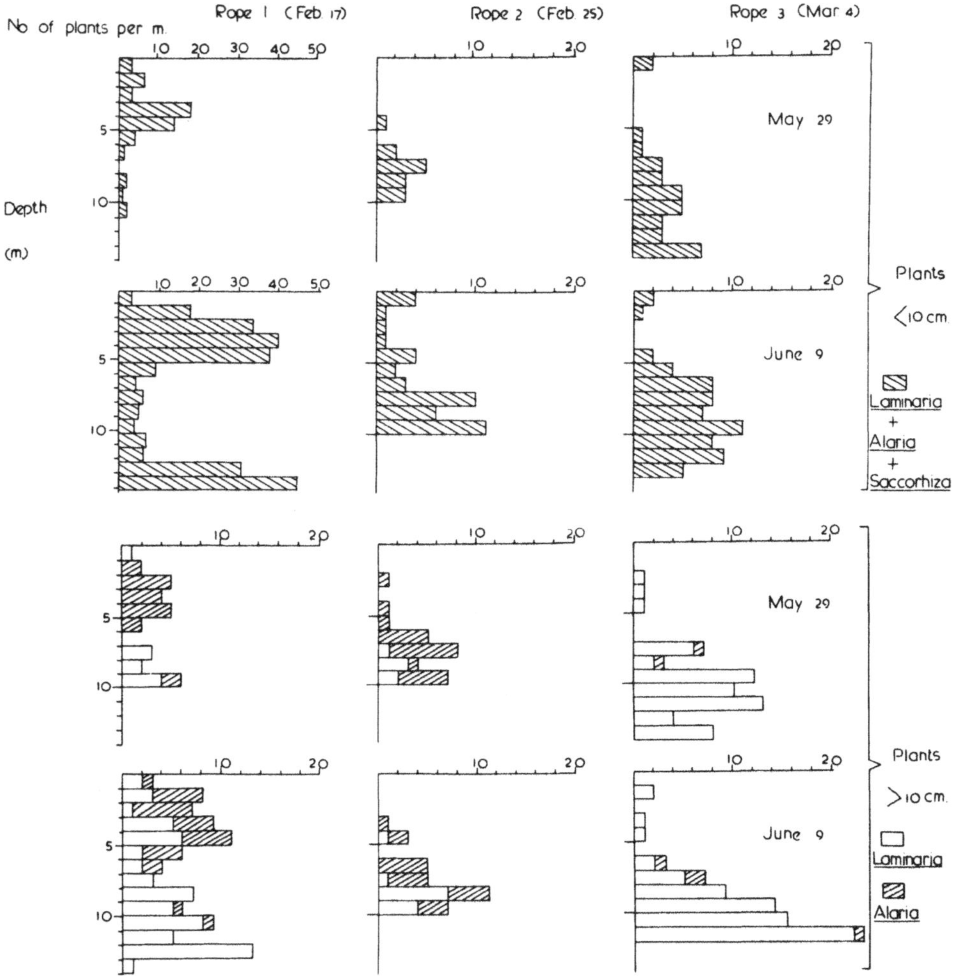

Fig. 1. Number of plants per metre found on vertical ropes placed in the sea in February and March. Total numbers of Alaria, Saccorhiza and Laminaria saccharina smaller than 10 cm, and numbers of Alaria and Laminaria longer than 10 cm are shown as observed on May 29th and June 9th.

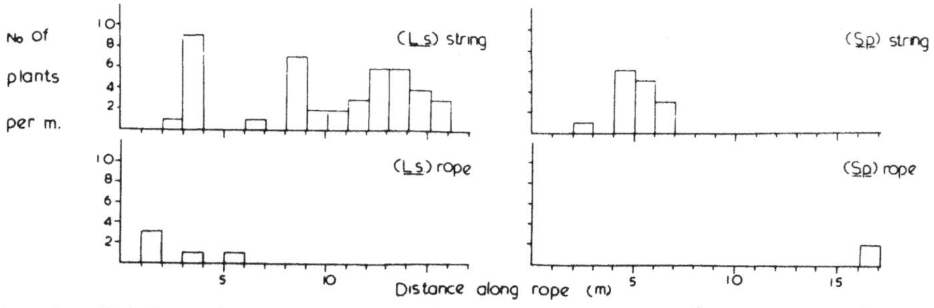

Fig. 2. Total numbers per metre of young sporophytes (unidentified) found on the horizontal seed string and rope after 2 months in the sea.

chosen so as to maximise energy production. For this purpose samples of Saccorhiza are being collected and dried monthly and sent to Dr Jensen in Trondheim for chemical analysis.

1.2.5 Drying

A few preliminary drying experiments have been performed by hanging the plants in still air at 29°C and a relative humidity of 50%. It would appear that all but the very largest specimens of Saccorhiza lose 99% of their water content within 48 hours, while Laminaria saccharina would probably only require 24 hours. The laminae dried faster than stipes and holdfasts. More drying experiments will be carried out in Nottingham.

2. Conclusions

A method of supporting single horizontal ropes at variable depths has proved successful but a more robust anchoring system will probably be required. The first artificial seeding technique was unsuccessful because conditions were inadequately controlled. Natural seeding of the vertical supporting ropes has been substantial, Alaria esculenta was an important colonizer in February and, because it is a low temperature species, suggest itself as an early crop. Saccorhiza polyschides, on the other hand, being a relatively high temperature species, should be seeded later. Plants of Laminaria and Saccorhiza should be easy to dry at slightly elevated temperatures.

References

1) Sundene, O. (1961) The implications of the transplant and culture experiments on the growth and distribution of Alaria esculenta. Nytt Magasin for Botanik., 9, 155 - 174.

2) Black, W.A.P. (1948) The seasonal variation in chemical constitution of some of the sublittoral seaweeds common to Scotland. III Laminari saccharina and Saccorhiza bulbosa. Journal of soc. chem. industries London, 67, 172 - 176.

3) Haug, A. and Jensen, A. (1954) Seasonal variations in the chemical composition of Alaria esculenta, Laminaria saccharina, Laminaria hyperborea and Laminaria digitata from Northern Norway. Report No. 4 of the Norwegian Institute of Seaweed Research.

4) Chynoweth, D.P., Klass, D.L. and Sambhunath, G. (1978) Biomethanatio of Giant Brown Kelp Macrocystis pyrifera. Institute of Gas Technolog Energy from Biomass and Wastes. Symposium papers Washington D.C. August 1978.

BIOMASS FROM OFFSHORE SEA AREAS

Contract Number	: ESE-R-021-UK*	
Duration	: 33 months	1 Oct 1980 - 30 June 1983
Total Budget	: £ 133,825	CEC Contribution : £ 60,000
Head of Project	: Professor J.G. Morley, University of Nottingham	
Contractor	: University of Nottingham, (Wolfson Institute of Interfacial Technology)	
Address	: University Park Nottingham NG7 2RD England	

Summary

Studies are being made of the feasibility of growing marine biomass on polypropylene ropes as a source of renewable energy and estimates of the total scale of this resource are given. Observations have been made of the drag forces experienced by polypropylene ropes when aligned transversely and longitudinally to the direction of water flow. Studies have also been made of the drag forces developed by simulated L. saccharina plants of various sizes in currents of various velocities. The indications are that the forces likely to be developed in weed covered ropes by tidal currents could be withstood by ropes which are sufficiently inexpensive to offer a reasonable prospect that this concept of a renewable energy source is commercially viable. These preliminary observations of course require confirmation by studies of the drag forces developed in ropes covered by an overgrowth of real plants.

*See also report from Liverpool University (joint contract).

1.1. Introduction.

The area of land available for the growth of biomass as an energy source is limited so it is of interest to assess the feasibility of the growth of marine biomass for this purpose. The objectives of this project are to provide an assessment of the possible scale of this resource and the technical and commercial feasibility of utilising it. Existing data indicates that marine biomass growth rates of approximately 2 kg dry weight per square metre per year are achievable. Initially it is being assumed that the calorific value of dry marine biomass will be about one half that of the equivalent weight of coal so that the potential output of 1 km^2 of sea surface per year is of the order of 10^3 tonnes of coal equivalent. Artificial structures on which to grow the biomass will be required. Such structures should lend themselves to easy harvesting and be sufficiently inexpensive and durable to enable the operation to be commercially viable. At present polypropylene rope seems the most promising substrate material. Marine biomass can be grown on rope, growth rates in the region of 15 kg fresh weight (approx. 2 kg dry weight) per metre of rope being observed near the Isle of Man (1). Preliminary estimates indicate that suitable moored ropes would cost of the order of £0.1 per metre, have a ten year life and produce biomass having a value of £0.3 (at present prices) per metre over the life of the rope. The rope would have to withstand forces generated by wave action, by tidal currents and also by the harvesting procedures used. Initially, it is being assumed that forces due to wave action will be eliminated by arranging for the total length of the rope to be greater than the distance between the mooring positions at each end. In this way, at any instant, part of the rope will be carrying small tensile loads due to wave action and part will be deformed in compression due to wave action. It is envisaged that harvesting would be carried out using boats. Two possible approaches for utilising the energy in marine biomass seem possible. The first would depend on anaerobic digestion to produce methane (2). The alternative approach would be to utilise power station waste heat to dry fresh biomass. It has already been shown (3) that seaweed can be dried to a water content of approximately 1% in 48 hours at 30 °C in air having a relative humidity of 50%. The preliminary investigations described below deal with the forces likely to be generated on seaweed covered ropes with steady water flow rates, e.g. tidal currents. It has been necessary to construct samples of "model" seaweed since fully grown plants are not yet available. The models have been constructed from polythene tubes and sheet having various dimensions comparable to single plants of L: saccharina. The density of the models has been made equal to the surrounding water. It is recognised that actual plants are slightly denser than the surrounding water, by an amount not yet known, and that this factor will affect the forces developed. However, it was felt that the data obtained would enable preliminary estimates of drag forces to be made.

1.2. Description of Apparatus and Measuring Equipment.

A simple horizontal water trough 2 m long 30 cms wide and 30 cms deep was used. A large electric pump passed water through the trough and the water then returned to the pump via two storage tanks arranged in series. Water flow rates were measured locally within the trough and average flow rates were obtained by measuring the rate of increase of the volume of water in the first storage tank after interupting the flow to the second. A probe was used to measure the drag forces. This was mounted above the tank extended below the water surface. Experimental samples were attached to the lower end of the probe. Measurements of drag forces were made on various lengths of polypropylene rope arranged longitudinally and transversely to the flow rate. Also drag forces were measured for various "models" of L: saccharina plants.

1.3. Results.

The drag forces developed per metre of rope aligned parallel with the direction of flow are shown in Figure 1 for various flow rates. The effects of the ends of the rope and the drag on the probe itself have been eliminated from the data presented. Over the range of flow velocities investigated, the drag forces are proportional to the square of the speed of the water flow. Much higher drag forces are developed when the rope is aligned transversely to the direction of water flow. Data for this condition is shown in Figure 2. Again the observed drag forces are proportional to the square of the velocity of the water flow.

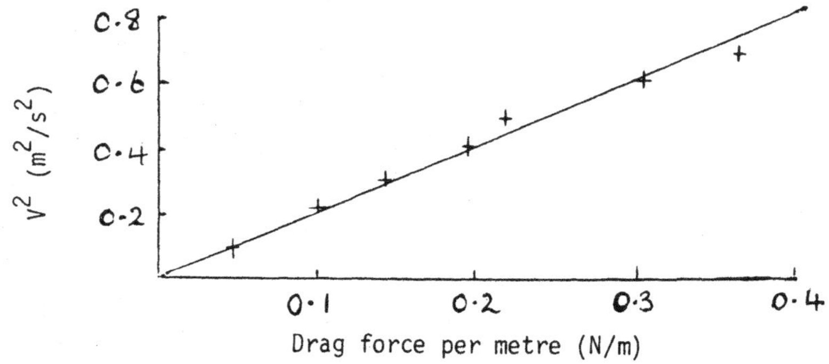

FIGURE 1 - Rope aligned in direction of flow

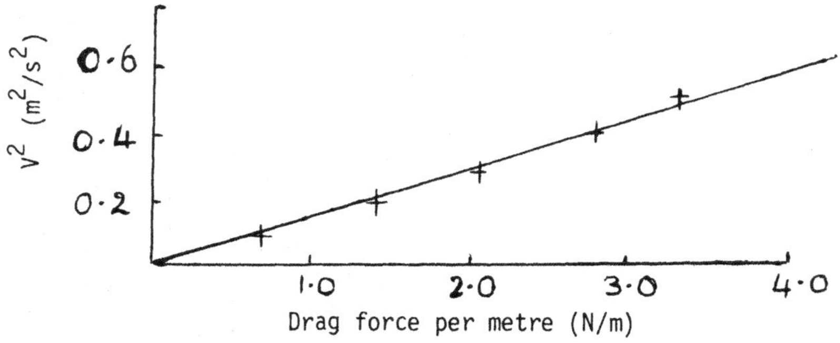

FIGURE 2 - Rope perpendicular to direction of flow

In Figure 3 the observed drag forces generated by various specimens of simulated seaweed are shown. The drag forces are again roughly proportional to the square of the current velocity and at low current velocities are also approximately proportional to the area of the artificial seaweed. However, in the case of some of the larger specimens the drag forces do not increase as rapidly with flow rate as would be predicted from this simple relationship between drag and water flow velocity.

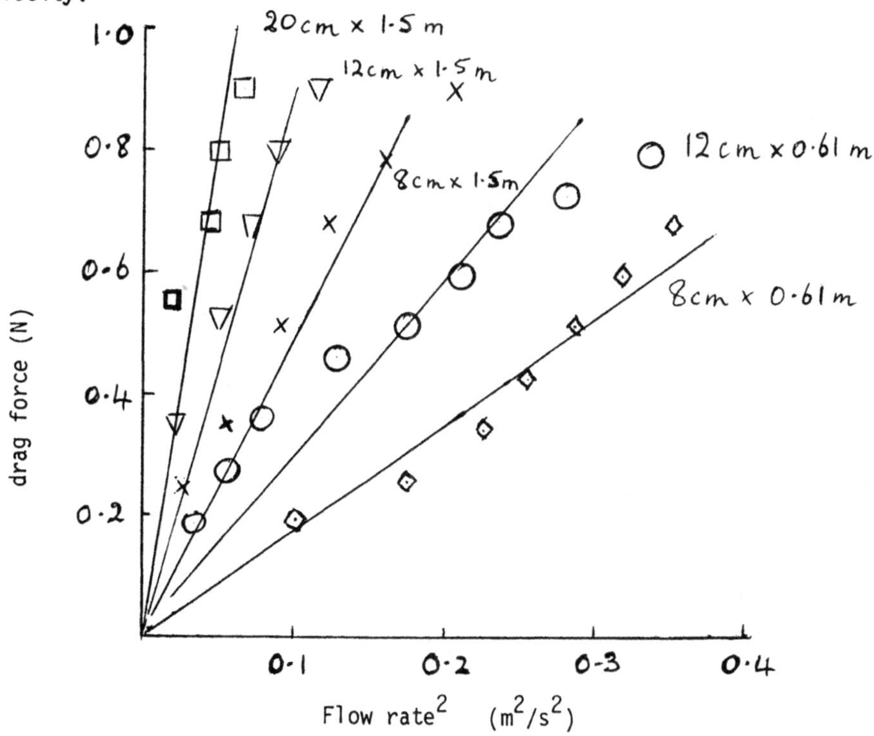

FIGURE 3 - Drag forces at various flow rates on simulated sea weed of various sizes

1.4. Analysis of Results and Comments.

According to simple theory the drag forces generated on a length of rope positioned transversely to the direction of flow will be given by

$$C_D = 2F/\ell V^2 A \quad (1)$$

Where F is the drag force V the velocity of the current ℓ is the density of the fluid (1000 kg/m^3) and A = length times diameter of the rope (1.4 cms). The drag coefficient C_D would be expected to be about unity and for this value the calculated drag force agrees well with equation (1). The data obtained to date can be used only as a rough guide in predicting the hydrodynamic characteristics of weed covered ropes because there are uncertainties about the validity of the model systems used. Since the currents reverse direction in a cyclic manner, it is necessary to consider the forces developed in the rope when the direction of current flow makes any angle θ with a line joining the two moored ends of the rope. Also the current velocity will not normally be constant at all angles of θ and this will depend on the actual geographical location considered. The forces developed in the rope for steady flow conditions can be calculated from suspension bridge theory. The maximum tensile stress in the rope will be developed at one or both ends. Its value can be calculated for all values of θ using suspension bridge theory. If the direction of current flow is parallel to the rope the maximum tensile load carried by the rope is developed at the upstream end. The limiting condition is obtained if it is assumed that all of the drag forces are supported by one end of the rope. These can be estimated by simply summing the drag forces developed by the weed over the whole length of the rope. An estimate of this can be made from the data already obtained as follows. Assuming a current velocity of 0.225 m/s (i.e. ∿ ½ knot) the drag developed by a single artificial plant of typical size (125 cms x 15 cms) will be about 0.5 N. If we assume 30 plants per metre of rope the drag per metre will be approximately 15 N, so that for a rope 1000 m long the force developed by a current of this velocity will be approximately 1.5 tonnes. This is within the capacity of the ropes currently in use in the experimental studies. If the same drag forces are assumed to be applicable, the maximum stress developed under transverse flow can be held below these values by a suitable choice of rope geometry.

2. Conclusions.

A production rate of the order of 10^3 tonnes of coal equivalent per annum may be attainable for 1 km^2 of sea surface so that a significant fraction of the energy requirements of Western Europe might be supplied by marine biomass using for this purpose a relatively small portion of the adjacent sea area. As a first step it is necessary to design sufficiently inexpensive moored artificial structures on which the seaweed would be grown. The cost of the structures will depend largely on the forces which they will have to withstand. Attempts are being made to avoid most of the effects of wave motion by designing compliant linear buoyant structures using polypropylene rope. Preliminary estimates have been made using simulated seaweed of the scale of drag forces likely to be generated by tidal currents. On the basis of this data it seems that the forces generated by modest tidal current (∿ 0.5 knot) would be acceptable. It will however be necessary to confirm this data using ropes covered with a seaweed overgrowth and the effect of breaking waves has still to be taken into account.

References.

1. Dr. J.M. Jones and T.J. Holt, (Private Communication).
2. R.A. Troiano, D.L. Wise, D.C. Augustein, R.G. Kispert and C.L. Cooney, Resource Recovery and Conservation 2, (1976), p. 171-176.
3. Dr. J.M. Jones and T.J.Holt, (Private Communication).

RENEWABLE HYDROCARBON PRODUCTION BY CULTIVATION OF THE GREEN ALGA *BOTRYOCOCCUS BRAUNII*

INVESTIGATION OF THE FACTORS AFFECTING HYDROCARBON PRODUCTION

Contract Number : ESE-R-022-F(S.D.).

Duration : 36 months 1 January 1981 - 31 December 1983

Total Budget : F. 778 720 contribution C.E. 30 %.

Head of Project : Dr. E. CASADEVALL (Directeur de recherche CNRS)

Contractor : Ecole Nationale Supérieure de Chimie de Paris.

Address : Laboratoire de Chimie Bioorganique et Organique Physique
E.N.S.C.P.
11, rue P. et M. Curie
75231 PARIS CEDEX 05

Summary

The green unicellular alga Botryococcus braunii (B.b.) with à hydrocarbon content as large as 75 % of the dry weight, hold the first place among biomass with a high energy content. The large scale cultivation of this alga.if economical profitability was reached, should allow to replace partly fossil hydrocarbons by renewable hydrocarbons.

So as to assess the optimum growth conditions for a large scale cultivation we have undertaken a systematic investigation on the weight of biological and physico-chemical culture parameters on biomass and hydrocarbon production.

The preliminar results, we have obtained show that the alga strain source, the species of the contaminating bacteria,the illumination duration, the concentration of dissolved gaz (O_2 , CO_2) and of some mineral nutrients (NO_2^-, PO_4^{3}) are of great importance both on culture growth and hydrocarbon content. It appears also that suitable adjustments of these parameters allow an important improvement of hydrocarbon productivity.

Valuable information on B.b. physiology and metabolism was gained through study of the variation in hydrocarbon content along growth curve, kinetics of dissolved salt absorption by the cells and ultrastructural investigation.

PRODUCTION RENOUVELABLE D'HYDROCARBURES PAR CULTURE DE L'ALGUE VERTE *BOTRYOCOCCUS BRAUNII*

Numéro de contrat : ESE-R-022-F(S.D.).

Durée : 36 mois à partir du 01/81.

Budget total : F.778 720 contribution C.E. 30 %.

Chef de projet : Dr. E. CASADEVALL (Directeur de recherche CNRS).

Contractant : Ecole Nationale Supérieure de Chimie de Paris.

Adresse : Laboratoire de Chimie Bioorganique et Organique Physique
E.N.S.C.P.
11, rue P. et M. Curie
75231 PARIS CEDEX 05

Résumé

L'algue verte unicellulaire Botryococcus braunii (B.b.) avec un contenu en hydrocarbures pouvant atteindre jusqu'à 75 % du poids sec se place au premier rang des biomasses à haut contenu énergétique.

La culture industrielle de cette algue, si elle était économiquement rentable pourrait permettre de remplacer, au moins en partie, les hydrocarbures fossiles par des hydrocarbures renouvelables.

Pour pouvoir apprécier la rentabilité d'une telle culture nous avons entrepris d'étudier l'influence des paramètres biologiques et physico-chimiques des cultures sur la production de biomasse et d'hydrocarbures. Les résultats encore très partiels obtenus montrent que l'origine de la souche et la nature des bactéries contaminantes, d'une part, la durée de l'éclairement, la concentration des gaz (O_2-CO_2) et des sels dissous d'autre part, ont une influence très grande sur la production de biomasse et du contenu en hydrocarbures. Le tracé de la courbe de croissance, la mesure de la variation des nitrates et des phosphates dans le milieu de culture tout le long de cette courbe, l'examen cytologique des cellules et l'analyse des hydrocarbures à différents stades du développement de la culture nous ont permis par ailleurs d'obtenir des informations intéressantes relatives à la physiologie et au métabolisme de B.b.

Introduction.

L'algue verte unicellulaire Botryococcus braunii (B.b.) présente la particularité jusqu'içi unique parmi les autres organismes photosynthétiques examinés, d'avoir un contenu en hydrocarbures pouvant atteindre 75 % de la biomasse sèche. Elle accomplit aussi la bioconversion directe de l'énergie solaire en un produit chimiquement identique aux pétroles fossiles et susceptible de les remplacer dans toutes leurs applications. La culture industrielle de B.b. pourrait donc être envisagée comme une voie possible pour l'obtention d'hydrocarbures renouvelables pouvant se substituer aux hydrocarbures fossiles.

On sait depuis quelques années que B.b. est impliquée dans la génèse des pétroles. Elle a donné lieu, à ce titre, à un certain nombre de publications. Cependant on trouve dans cette littérature assez peu de données relatives à la physiologie et au métabolisme de cette algue, ainsi qu'aux conditions de sa culture. Or ces données sont indispensables pour tester la faisabilité d'une production renouvelable d'hydrocarbures à partir de sa culture. C'est pourquoi le contrat qui a été établi entre notre laboratoire et la C.E. est principalement orienté vers l'étude de l'influence des paramètres de culture sur la production de la biomasse et des hydrocarbures ainsi que sur celle des caractères métaboliques et physiologiques de B.b.

Nous rapportons ci-dessous les premiers résultats de cette étude :

Conditions de culture.

Nous avons utilisé deux types de cultures pour lesquelles les conditions étaient les suivantes :

Culture en erlen fermé.

milieu Chu 13 x 2 - Eclairement jour-nuit 14/10 ; t° = 20°C - sans agitation.

Culture en air-lift.

Conditions I - milieu Chu 13 x 2 ; air enrichi CO_2 1 % - Eclairement continu ; t° = 25°C.

Conditions II - milieu Chu 13 x 2 ; air enrichi CO_2 4 % - Eclairement jour-nuit 14/10 ; t° = 25°C.

Résultats.

L'étude de paramètres physico-chimiques et biologiques susceptibles d'influer sur le développement des cultures de B.b. et sur la production en hydrocarbures nous ont permis d'obtenir les résultats rapportés dans les tableaux suivants :

TABLEAU I - Influence des conditions de culture et de quelques paramètres physico-chimiques (durée de l'éclairement, concentration des gaz et sels dissous) sur la croissance et la production d'hydrocarbures chez B.b.

n° 1 - culture en erlen fermé, durée de la culture 30 j.
n° 2 - culture en air-lift, conditions I, durée de la culture 21 j.
n° 3 - culture continue, renouvellement total du milieu de culture après 4 jours par apport d'un égal volume de milieu neuf.
n° 4 - culture continue, comme 3, mais renouvellement en 2 j.
n° 5 - culture air-lift, conditions I, prélèvement après 17 j.
n° 6 - comme n° 5, mais concentration initiale (NO_3^-) x 2.

Ces cultures, sauf n° 1, ont été réalisées en collaboration avec C. GUDIN et C. CHAUMONT, SF-BP Lavéra. L'échantillon mis en culture provenait de l'algothèque de Gottingen.

(suite du tableau I page suivante)

Tableau I (suite)

Essais N°	Poids sec (g/l)	Teneur de la culture en hydrocarbures (g/l)	Concentration en hydrocarbures (en % du poids sec)	Productivité en hydrocarbures (g/l/jour)
1	2	0,34	17	0,011
2	6	1,77	29,5	0,084
3	1,33	0,21	16	0,053
4	0,71	0,10	14	0,050
5	5,3	1,50	29	0,088
6	7	2,52	36	0,148

Les valeurs de productivité ainsi exprimées peuvent paraître faibles; cependant une culture à grande échelle ayant par exemple la productivité n° 2 fournirait 60 tonnes d'hydrocarbures/ha/an pour une épaisseur d'eau de 20 cm.

TABLEAU II - Productivité d'une culture de B.b. (échantillon Gottingen, air-lift, conditions I) en fonction de l'âge de la culture.

Prélèvement n°	Temps de culture (g)	Poids sec biomasse (g/l)	Poids hydrocarb. (g/l)	% hyd. / Pds biomasse	Production hyd. (g/l/j (1)
1	3	0,6	0,216	36,4	0,072
2	5	1,1	0,433	38,4	0,108
3	7	2,8	0,958	36,4	0,260
4	14	4,5	1,128	23,5	0,024
5	21	4,9	1,268	26,0	0,020
6	31	4,3	0,801	18,8	négatif

(1) La productivité est calculée pour l'intervalle de temps compris entre deux prélèvements successifs.

TABLEAU III - Variation en fonction du temps de culture de la concentration en nitrate, nitrite et phosphate du milieu pour une culture (échantillon Gottingen) air lift, conditions I.

Echantillon n°	Temps de culture en jours	(NO_3^-) mg/l	(NO_2^-) mg/l	(PO_4^{3-})	pH
0	0	120,5	traces	19,7	7,6
1	3	81,6	"	traces	8,1
2	5	83,8	"	"	7,9
3	7	83,8	"	"	8,6
4	14	44,8	"	34,3	7,4
5	21	33,4	"	35,0	7,2
6	31	63,3	"	22,3	7,7

TABLEAU IV - Influence de l'origine de l'échantillon d'algue sur la production de biomasse et d'hydrocarbures pour une culture de B.b., air lift, conditions II.

Origine	Temps de culture en jour	Hyd. mg/l	% Hyd. / biomasse	Productivité hyd. g/l/jour
Gottingen	0 (1)	56	22,9	-
"	9 (1)	178	17	0,014
	14 (1)	170	13	0,008
	18 (2)	269	14	0,012
Thonon	0 (1)	20	8,7	-
"	9 (1)	131	10,2	0,012
"	14 (1)	193	12,1	0,012
"	18 (2)	233	11,0	0,012
Cambridge	14 (2)	5	2,5	0,0004

(1) hydrocarbures totaux.
(2) hydrocarbures externes.

TABLEAU V - Influence du type de bactérie contaminante sur la production de biomasse et d'hydrocarbures par les cultures de B.b. - Cultures en erlen. Témoin = algue Thonon contaminée par une bactérie du genre Nocardia.

Espèce contaminante	Psd 1	Psd 2		Psd 3		Psd 6
Durée de la culture en jours	50	38	114	41	111	47
Pds ext.hex. mg/l (1)	91	70	220	69	248	220
Pds biomasse mg/l	970	260	1470	750	1360	1560
% ext.hex/biomasse	9,4	26,5	14,9	9,1	18,1	14,1

Espèce contaminante	Psd 7	Arthrob.		Coryn.	Témoin
Durée de la culture en jours	42	39	115	38	40
Pds ext.hex. mg/l (1)	115	310	148	133	78
Pds biomasse mg/l	940	1070	1120	1310	920
% ext.hex/biomasse	12,3	29,1	13,5	10,2	8,5

(1) L'extrait hexanique contient les hydrocarbures externes de B.b. (les hydrocarbures internes ne sont pas extraits dans ces conditions il contient également les hydrocarbures bactériens et quelques autres composés lipidiques (esters d'acides gras). Des expériences antérieures nous ont montré que ces deux dernières fractions constituent environ 30 % de la totalité de l'extrait hexanique.

Commentaires.

1) Influence des paramètres physico-chimiques.

B.b. est décrite dans la littérature comme une algue à croissante lente. D'ailleurs dans des conditions de culture en erlen (tableau I, essai 1) nous avons observé des temps de doublement de la biomasse de 1 semaine environ. Cependant si l'on compare entre elles les données du tableau I, on observe qu'à de faibles variations des conditions de culture correspondent de fortes variations de la productivité en hydrocarbures. En effet, on note (essais 1 et 2) qu'une augmentation de la durée de l'éclairement ainsi qu'un enrichissement en CO_2 sont favorables à la production d'hydrocarbures. Il en est de même pour l'augmentation de la concentration en nitrate (essais 5 et 6). Les essais 3 et 4 montrent qu'en culture continue, pour des doublements de biomasse de 2 jours, la productivité en hydrocarbures reste encore élevée. La production d'hydrocarbures pour ces deux derniers essais, à partir d'une culture en développement rapide, semble indiquer que la biosynthèse des hydrocarbures par B.b. est un effet du métabolisme normal de cette algue.

2) Influence des paramètres biologiques.

Courbe de croissance. A partir de conditions de culture (essai 2 du tableau I) assurant un développement rapide de l'algue, nous avons pu établir une courbe de croissance sur la base de la variation, en fonction du temps, du poids de biomasse produite (tableau II). Dans la première partie de cette courbe nous observons une croissance exponentielle qui correspond à des temps de doublement de 2,5 jours, puis un ralentissement et finalement une phase de repos où la quantité de biomasse et d'hydrocarbures précédemment formés diminuent.

La productivité en hydrocarbure déterminée en fonction de l'âge de la culture, fait apparaître un maximum qui se situe sensiblement au milieu de la phase exponentielle. La productivité baisse ensuite et devient même négative en fin de culture. Cette baisse de productivité peut avoir au moins trois origines différentes. Elle peut résulter soit d'une dégradation du matériel cellulaire, soit d'une modification du milieu de culture (appauvissement en certains constituants) soit d'une dégradation des hydrocarbures par les bactéries contaminant la culture. Nous avons en effet montré dans des travaux antérieurs que B.b. est incapable de réutiliser les hydrocarbures qu'elle produit.

Morphologie cellulaire. Afin de visualiser les modifications de morphologie cellulaire susceptibles de se manifester aux différents stades de la culture, nous avons effectué un examen cytologique des cellules pour différents temps de culture. Les observations qui ont été faites peuvent se résumer de la façon suivante : En début de culture (tableau II, prélèvement 1) on observe la présence d'hydrocarbures externes sous la forme de petits globules. Il y a peu de vacuoles (hydrocarbures internes) et peu de grains d'amidon stockés dans le chloroplaste. Au 2ème prélèvement les globules externes et les vacuoles ont grossi, les grains d'amidon sont plus nombreux. Au 3ème prélèvement l'amidon a envahi le chloroplaste, les vacuoles sont plus grandes, les globules plus nombreux et très anastomosés. Dès le 4ème prélèvement les contenus cellulaires sont dégénérescents. Pour les prélèvement suivants cette dégénérescence s'accentue et la quantité d'hydrocarbure semble avoir cessée de croître. Ceci est effectivement confirmé par le dosage des hydrocarbures (tableau II). Cette étude cytologique montre que dès le milieu de la phase exponentielle (prélèvement 3) on n'a plus que des cellules qui ont cessé de se multiplier. L'augmentation de biomasse que l'on observe résulterait donc seulement de l'augmentation de la taille des cellules et de la quantité d'hydrocarbures. L'aptitude à produire des hydrocarbures se poursuivrait jusqu'à cessation de toute activité cellulaire.

Evolution du milieu de culture. Afin de déterminer si les différents stades de la croissance sont liés à des modifications de la composition du milieu de culture, nous avons mesuré la variation en fonction du temps de la concentration en nitrates, en nitrites (leur présence pouvant être considérée comme un facteur d'inhibition de la croissance) en phosphates, ainsi que la variation du pH. Les résultats de ces mesures sont rapportés sur le tableau III. On constate que le pH varie peu (les deux pics enregistrés à 8,1 et 8,6 pourraient correspondre à des blooms bactériens). On n'observe pas un épuisement total du milieu en nitrates alors que s'installe vers le 17ème jour la phase de repos. Il n'apparait pratiquement pas de nitrites. L'évolution de la concentration en phosphates est assez remarquable. En effet ceux-ci disparaissent dès le troisième jour, début de la phase exponentielle, pour réapparaître lorsque la culture a pratiquement atteint la phase de repos. Cette disparition des phosphates du milieu ne limite pas la croissance, puisque la culture continue à se développer longtemps après que les phosphates aient disparu. Ce résultat semble indiquer que les phosphates sont stockés à l'intérieur de la cellule pour être ensuite réutilisés au fur et à mesure des besoins. L'examen en microscopie électronique des cellules 2 heures après leur introduction dans un milieu riche en phosphates montre la présence dans les vacuoles de petits grains osmiophiles. Ceux-ci pourraient être des amas de polyphosphates. Des expériences seront entreprises à l'aide de phosphates marqués afin de préciser la nature et la localisation d'un éventuel stockage intracellulaire.

Enfin, dans le but de déterminer la concentration initiale en phosphate minimale qui est nécessaire au bon développement des cultures dans les conditions utilisées, nous avons étudié l'influence sur la courbe de croissance d'une variation de cette concentration initiale. Pour cela nous avons réalisé des cultures avec des concentrations $(PO_4^{3-})_o$ multipliées par un facteur 2, 1, 0.5, 0.2 et 0.1. Pour les trois premières valeurs les courbes de croissance se superposent à peu près exactement. On observe par contre un léger ralentissement de croissance pour les deux dernières. La concentration $(PO_4)_o$ limitante pourrait donc se situer entre $(PO_4)_o$ x 0,5 et $(PO_4)_o$ x 0,2.

Origine de la souche - Nature des bactéries contaminantes.

Les échantillons d'algues dont nous disposions au début de cette étude provenaient de trois algothèques : Cambridge, Gottingen, Thonon. Ces trois échantillons ont une origine commune la souche LB-807-1. L'analyse bactériologique que nous avons effectuée sur ces échantillons a montré qu'ils diffèrent par les bactéries qui les contaminent. L'algue de Cambridge est contaminée par 18 espèces de bactéries différentes, alors que les algues de Thonon et de Gottingen sont l'une et l'autre associées à une seule espèce bactérienne. Les résultats de cultures effectuées dans les mêmes conditions à partir de ces trois échantillons, font appraître (tableau IV) que les algues de Thonon et de Gottingen conduisent à des poids de biomasse assez voisins et ont des productivités en hydrocarbure comparables. L'algue de Cambridge se distingue des deux premières par des valeurs beaucoup plus faibles, aussi bien pour la biomasse que pour les hydrocarbures produits. Par contre, l'analyse chromatographique des hydrocarbures extraits des trois échantillons montre que les composés prédominants dans ces extraits sont les mêmes. Il s'agit de quatre diènes en C_{25}, C_{27}, C_{29} et C_{31} ayant la formule générale $CH_3-(CH_2)_7-CH=CH-(CH_2)_n-CH=CH_2$ (avec n = 13, 15, 17, 19), et d'un triène en C_{31}. Les pourcentages relatifs de ces différents composés sont voisins pour les trois échantillons.

L'origine commune des échantillons d'algue étudiés et le fait qu'ils produisent tous les mêmes types d'hydrocarbures, suggéraient que seul le grand nombre de bactéries contaminant l'échantillon de Cambridge était à l'origine de sa faible productivité. Afin de préciser l'influence des bactéries contaminantes sur les cultures de B.b. nous avons procédé à deux séries d'expériences.

- d'une part, nous avons essayé de déterminer l'influence de chacune des espèces contaminant l'algue de Cambridge sur la production de biomasse et d'hydrocarbures.

-d'autre part, nous avons mis en culture successivement chacune des espèces bactériennes précédemment isolées des cultures d'algues sur un milieu contenant un hydrocarbure comme seule source de carbone.

Pour la première série d'expériences comme nous ne disposions pas d'une souche de B.b. axénique et que la purification de nos échantillons s'avérait difficile nous avons décidé de prendre pour culture de référence l'algue de Thonon à laquelle est associée une seule espèce de bactérie (Nocardia) et de la contaminer successivement par chacune des bactéries isolées de la culture de l'algue de Cambridge. Le poids de biomasse produite, ainsi que celui de l'extrait hexanique ont été déterminés pour les associations ainsi constituées. Les résultats de ces déterminations rapportés dans le tableau V montrent que les cultures associées à certaines bactéries (c'est notamment le cas des Pseudomonas 6 et 7, d'Arthrobacter et de Coryne-bactérium) produisent davantage de biomasse et fournissent plus d'extrait hexanique que le témoin. Par contre, la présence d'autres bactéries (Pseudomonas. 2 et 3) semble se traduire par de faibles rendements, aussi bien en biomasse qu'en extrait hexanique. Les résultats de la deuxième série d'expériences indiquent par ailleurs que la plupart des espèces bactériennes isolées des cultures peuvent croître lorsqu'un hydrocarbure est la seule source de carbone de la culture. Les deux espèces de Nocardia qui contaminent respectivement l'algue de Thonon et l'algue de Gottingen, se développent assez activement dans ces conditions.

Cette dernière observation permet d'interpréter la disparition d'une partie des hydrocarbures constatée (productivité négative du tableau II) en fin de culture pour l'algue de Gottingen, ainsi que la baisse des rapports extrait hexanique/biomasse lorsque le temps de culture passe de $\simeq$ 40 j à plus de 100 j. pour certaines des associations algue de Thonon-bactérie présentées dans le tableau V. On peut penser que dans l'un et l'autre cas la bactérie de l'espèce Nocardia qui contamine l'algue intervient pour dégrader une partie des hydrocarbures que celle-ci produit ;le bilan devenant négatif lorsque la culture d'algue vieillit et que la dégradation l'emporte sur la production. Ces résultats montrent de façon assez évidente l'influence néfaste des bactéries qui dégradent les hydrocarbures. Toutefois le caractère bénéfique (possibilité de l'apport aux cultures de facteurs de croissance) d'autres types de bactéries qui est suggéré par quatre résultats du tableau V devront être précisés en utilisant comme témoin une souche de B.b. axénique.

Conclusion.

Les résultats que nous avons obtenus concernant l'influence sur la production des hydrocarbures par B.b. de l'éclairement, de la concentration des éléments minéraux (NO_3^-, PO_4^{3}) et des gaz dissous (O_2, CO_2) dans le milieu de culture, ainsi que de l'origine de la souche et de la nature des bactéries contaminantes, sont encore très partiels. L'étude qui a été effectuée nous a néanmoins permis d'acquérir de nombreuses informations relatives à la physiologie et au métabolisme de cette algue. Elle a par ailleurs mis en évidence la nécessité d'une étude systématique et étendue de l'influence de ces paramètres, tant physico-chimiques que biologiques pour pouvoir définir des conditions optimales de culture qui permettraient de passer sans trop de risques de l'échelle du laboratoire à celle de "cultivateur" de grande capacité.

SUBJECT IV. BIOMASS CONVERSION (BIOLOGICAL ROUTES)

The anaerobic digestion of farm wastes and energy crops

Methane from agricultural wastes and from energy crops

Methane production by anaerobic digestion of algae

Two-phase process for the anaerobic digestion of organic wastes yielding methane and compost

Starting up of a thermophilic anaerobic digestion

Alcoholic fermentation - Improvement of the technology based on physiological phenomena

Utilization of agricultural wastes as energy sources - Enzymatic hydrolysis of cellulose to glucose

Liquefaction and saccharification of agricultural biomass

THE ANAEROBIC DIGESTION OF FARM WASTES AND ENERGY CROPS

Authors	: D.A. STAFFORD, D.E. HUGHES, S.P. ETHERIDGE
Contract Number	: ESE-R-023 UK
Duration	: July 1980 - June 1983
Total Budget	: 100 000 EUA, EEC Contribution 50 %
Head of Project	: Dr. D.A. Stafford Professor D.E. Hughes
Contractor	: University College, Cardiff
Address	: Dept. of Microbiology University College Newport Road Cardiff Wales U.K.
Sub Contractor	: None

Summary:

One of the major influences for the production of energy from farm wastes and energy crops is that of cost. This includes the capital and running costs as well as the way digestion influences the operational parameters on the farm, particularly with respect to slurry handling.

A number of digester designs are being evaluated with respect to installation problems, digester performance and cost effectiveness.

Energy crops are being grown and will be used in admixture with farm wastes(pig and cattle), to produce biogas for direct use as a heating fuel, or for conversion to electricity using a cogenerator set.

RESEARCH INTO THE DEVELOPMENT OF PROTOTYPE UNITS FOR THE PRODUCTION OF BIOGAS METHANE FROM ENERGY-CROPS AND FARM WASTES

An agricultural plot of about 1.5 ha. has been sown with a 50/50 mixture of Italian Rye Grass and Forage Pea (Late March sowing). The development of the crop has been good without the addition of artificial fertilizer and an expected yield of 30-50 tonnes of energy crop will be ensiled before feeding to the anaerobic digesters. Percentages of dry matter and organic material is being monitored during the crop growth cycle. Digestibility of the crop is also being monitored in laboratory-scale (10-1) anaerobic digesters.

The ensiled grass produced at the beginning of 1981 is being fed to the Plug Flow digester, recently commissioned and fairly good gas yeilds are being obtained,(0.3 m3 Biogas/ Kg. VS added).

Two digesters at the pilot-scale are now in operation and the following provides a summary of installation and operational producers:-

Plug Flow Digester

One type of design chosen for investigation at Cleppa Park is the "Plug Flow" digester (30 m3). The unit installed was designed at University College and built locally by a fibre glass manufacturer. The most appealing aspects of this design were its low cost, ease of installation and simple operating procedure.

The digester is an insulated fibre glass cylinder of 6.5 m length with round enclosed ends to form receptacles for influent and effluent. Running through the centre of the digester is a series of mild steel pipes which act as heat exchangers when hot water is passed through the system. Each half of the heat exchanger can be turned by chain and pulley and in this way it can also be used as a stirrer. Inclined plates attached to the pipes help to move the contents along the digester to enhance the plug flow effect.

Installation was complete within 3 days but this could be reduced to 1 day. The digester was lowered into a hole of approximately dimensions onto a sand bed inclined 6 inches so that the effluent and was lower than the influent end, (see Figure 1)). The digester was then surrounded by concrete and a sloping apron was constructed at each end. Plastic pipes were sunk into the digester surface at various points so that thermocouples could be used to study temperature profiles.

Problems encountered during commissioning included leaking of the digester contents. This occured through various cracks which appeared during curing of the concrete. This may have been due to the very wet conditions at the time. Various filling materials were tried but only heavily waterproofed concrete proved effective. As an extra precaution the whole concrete surface was treated with a waterproofing agent. The heat exchanger was also tested for leaks but none were found.

The digester is now in operation as indicated and a regime of 100% pig waste is being fed, which will change through a number of concentrations including mainly energy crop and a small percentage volume of pig waste. No operational difficulties have yet been encountered. The gas will be stored and used to power a cogenerator set.

Hydraulic Digester

This digester (2 m3) was also designed at University College and built locally. It consists of 2 cones, one inserted into the other. The upper cone is split into 2 holding tanks, the influent and effluent tanks. The lower come comprises of the contents tank and the gas collection space.

The unit is heated by a thermostatically controlled steam input set at 35°C. The thermostat is also connected to a recycle pump so that when steam is required the digester contents are recycled within the lower cone.

Each day the digester is fed by pumping pig slurry into the feed tank. From there is trickles into the contents tank and displaces the same volume of effluent. The gas is collected from the space above the contents and its release is controlled and monitored.

The conical shapes within the digester serve to concentrate the solids fraction and hence the effluent holding tank also acts as a settler.

Within the first 2 months of continuous operation, retention times of lower than 8 days were reached with successful gas evolution. Gas readings of more than 2.2 m3 were recorded at this retnetion time with a digester volume (contents tank) of approximately 2 m3.

Total solids measurements of the feed slurry varied considerably and this was due to variations at the supply point and settling within the main feed storage tank, external to the digester.

This digester is very effective at concentrating the solids fraction of the slurry (solids reduction between influent and effluent is about 90%) and hence yielding a greater gas volume but is more complex due to valve manipulation and maintenance of hydraulic levels.

Contact Digester

A conventional completely mixed digester (30 m3) is being built by Hamworthy Engineering Ltd, Poole, Dorset, UK and installation is expected in early July, 1981.

All 3 digester units will be monitored concurrently, each receiving similar feeding regimes. Performances of the 3 designs can then be evaluated for gas yields and other operational parameters.

Full-Scale Digester - Chediston

The 350 m3 digester, built by Hamworthy Engineering Ltd., has been operating successfully for 1 year and about 2,000 hours has been obtained on the cogenerator set producing 20 Kwh. The electricity usage is 100% and has a bearing on the digestion costs and benefits. Fertilizer and animal feeding trials are well underway to provide data on cost-savings for the farmer when using digester residues. This data will be provided and evaluated at the next Coordination meeting.

References

1) Stafford, D.A., Hawkes, D.L. and Horton, R. Methane production from waste organic matter. CRC Press, USA 1980

2) Stafford, D.A. Chemicals from waste. Applied Science Publishers (In Press) 1981.

3) Stafford, D.A.. Wheatly, B.I. and Hughes, D.E. Anaerobic Digestion, Applied Science Publishers, 1980.

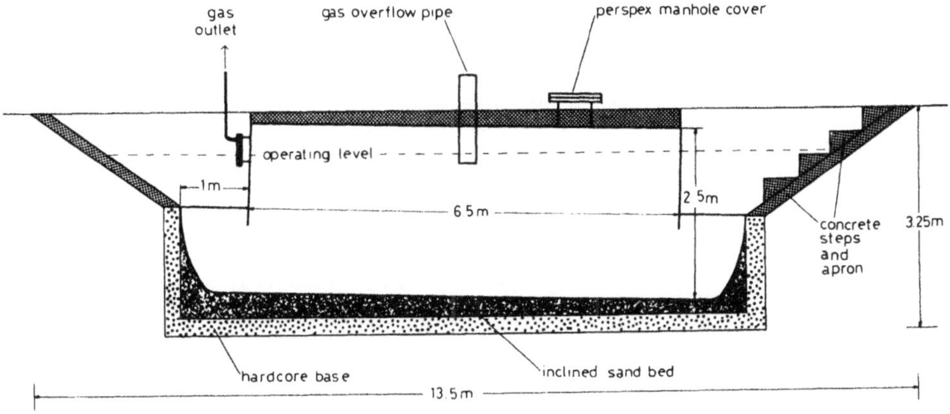

FIGURE 1

PLUG FLOW DIGESTER

METHANE FROM AGRICULTURAL WASTES AND FROM ENERGY CROPS

Contract Number	:	ESE-R-024-EIR	
Duration	:	36 months	1 July 1980 - 30 June, 1983
Total Budget	:	£133,050	CEC Contribution : £66,525
Head of Project	:	Dr. Emer Colleran, Department of Microbiology	
Contractor	:	University College, Galway.	
Address	:	Galway, Ireland.	

Summary

A two-stage process for the anaerobic digestion of solid agricultural wastes and energy crops was developed. The process utilizes a first stage reactor of simple design in which the hydrolysis and acidification phases of digestion are optimized. Liquefied organic material from the first stage reactor is removed by semi-continuous displacement with the effluent from a biogas reactor digesting pig slurry supernatant and is passed to the second stage fixed bed reactor for methanation. Approximately 50% of the total solids of wheat straw is liquefied during an 18-day holding period in the first stage reactor and converted to methane on an 80% COD reduction basis by a 3-day liquid retention cycle in the methane reactor. Methane yields in the order of 0.16 m^3 per Kg total solids are readily obtainable with chopped straw. Less refractory substrates give higher gas yields as indicated by a yield of 0.24 m^3 gas per Kg T.S. obtained in preliminary trials with a grass substrate. Scale-up of the fixed bed reactor to a 20 m^3 unit treating the liquid fraction of mechanically separated pig slurry was also attempted. The unit yielded results comparable to those obtained from high-rate conventional digesters when operated on a 10-day liquid retention time. Various design modifications are currently being assessed in order to decrease the liquid retention time to the 3-day cycle obtained in laboratory and pilot-scale units treating the supernatant, low solids fraction of gravity-settled slurry.

1.1 Introduction

The anaerobic digestion of complex substrates may be considered as a 'two-phase' biological process in which organic polymers are first metabolized by non-methanogenic microorganisms to volatile fatty acids which are then degraded by the methanogenic species to methane and CO_2 (1,2). The hydrolysis and acidification reactions of the first phase are carried out by facultative anaerobic species which differ significantly from the strictly anaerobic methanogens with respect to physiology, nutritional requirements, growth and metabolic characteristics, environmental optima and sensitivity to environmental stresses (3).

It is generally accepted that methanation is the rate-limiting step in the anaerobic digestion of non-polymeric substrates, such as glucose, in conventional single stage reactors (4). By contrast, the hydrolysis step becomes rate-limiting in the case of polymeric substrates such as cellulose (5). Given the major metabolic differences between the hydrolytic and methanogenic populations, many researchers now consider that successful digestion of polymeric substrates may best be achieved by a two-stage system in which the hydrolysis/acidification and the methanogenic phases are optimized in separate reactors (6).

This project is concerned with the development of such a two-stage system of digestion for agricultural residues such as straw and for purpose grown energy crops.

1.2 Apparatus

The hydrolysis/acidification phase was carried out in 1 L or 10 L batch reactors. Solid substrates, such as straw, grass, kelp, paper, etc. were added at varying concentrations to raw pig slurry so as to completely fill the reactor. The reactors were fitted with an inlet and outlet port and with a gas collection and sampling system.

A laboratory scale upflow fixed bed reactor of 9.2 L liquid capacity was utilized for the methanation phase. The reactor was constructed from clear perspex piping fitted with wavin end-pieces and contained flocor plastic rings as the support matrix. Acidified liquor from the first-stage reactor was pumped upwards through the matrix bed. The reactor was fitted with a gas sampling port and gas output was measured by a water-displacement system.

Scale-up of the fixed bed reactor from laboratory to pilot has already been reported (7,8). A full-scale plant was commissioned to the design outlined in Fig. 1.

In order to compare the results obtained with a two-stage system of cellulosic substrate digestion, a single stage 'solids' digester was also constructed. This consisted of a stainless steel tank of approximately 250 L capacity. The digester was insulated with expanded polyurethane foam and fitted with inlet and outlet liquid ports, a gas sampling and gas collection system and a series of temperature probes.

1.3 Results

Initial studies on the batch hydrolysis reactor using straw/pig slurry feeds indicated that rapid solubilization and fermentation of cellulose and hemicellulose occurred during the initial 8-10 days period. Thereafter, hydrolysis rates and volatile fatty acid production declined rapidly and very little further solubilization occurred until the onset of methanogenesis at day 14-18. The initial rapid rate of solubilization of the digestible fraction of straw was shown to be maintained by a liquid displacement regime in which the liquid fraction of the hydrolysis reactor was withdrawn every 5-6 days and replaced by an equal volume of effluent from a fixed bed methanation reactor treating the supernatant fraction of settled pig slurry (Fig. 2). Such a replacement system also prevented the onset of methanogenesis thereby ensuring separation of the hydrolysis/acidification and the methanation phases.

Preliminary studies on unseeded phase 1 reactors revealed a short lag phase which could be overcome by seeding with a cellulolytic bacterial population cultured on cellulose/slurry substrates. The addition of polymer degrading enzymes or commercial microbial preparations such as Polybac increased the initial rate of solubilization in unseeded reactors but had no effect on either the rate or the extent of solubilization in seeded systems. The effect of fine chopping and alkali pretreatment on the solubilization of straw in phase 1 reactors was also investigated. Table 1 indicates some of the results obtained.

TABLE 1. Solubilization of Cellulosic Substrates in a Batch Hydrolysis Reactor

Substrate	Pretreatment	Time (days)	% Total Solids Removed	% Cellulose Solubilized
Straw	Roughly chopped (1-2 cm)	18	41-50	61-81
Straw	Finely chopped (0.1 - 0.5 cm)	18	26-32	40-46
Straw	Ball-milled	18	40	61
Straw	Rough : ammoniated	18	41	66
Straw	Rough : NaOH treated	18	73-85	69-75
Grass	Lawn-mowings	8	70	75-80

Preliminary experiments on methanation of the acidified liquor from the phase 1 hydrolysis reactors indicate that COD removal rates of 80-90% are achievable using a liquid retention time of 3 days. Gas yields are in the order of 0.37 L CH_4 per gramme COD removed and the percentage CH_4 in the biogas ranges between 80 and 86%. In practice, the fixed bed reactor is not continuously linked to the phase 1 reactor and is fed on a daily basis with the supernatant fraction of settled pig slurry. The acidified liquor from the phase 1 reactor is simply passed to the fixed bed reactor as it becomes available from the liquid displacement cycle in operation.

In the alternative single stage solids reactor, unchopped straw (16 Kg) was mixed with cattle and pig slurry to a final moisture content of 64% and a density of 130-150 Kg/m^3. An initial self-heating phase of 5-6 days was allowed to take place by aeration of the reactor at defined intervals. Temperatures of 60-70°C were readily obtained during this period. After 6 days, the reactor was completely filled with pig slurry,

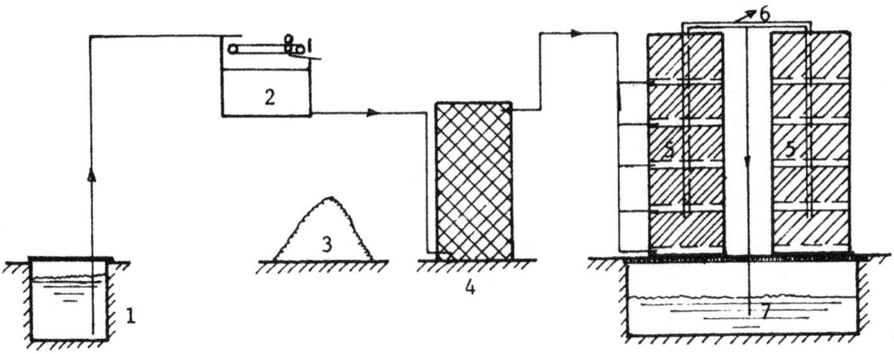

1. Holding sump.
2. Solids liquid separator.
3. Separated solids.
4. Heat exchanger.
5. Anaerobic digester.
6. Biogas.
7. Treated effluent.

Fig. 1. Schematic diagram of full-scale fixed bed anaerobic reactor.

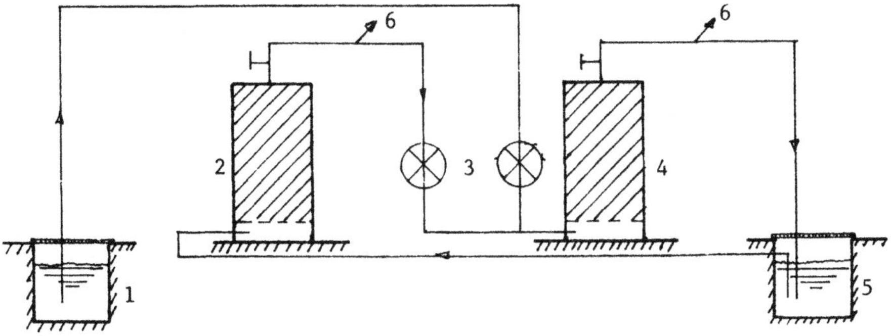

1. Raw slurry supernatant holding tank.
2. Hydrolysis reactor.
3. Valves.
4. Methanation reactor.
5. Effluent holding tank.
6. Biogas.

Fig. 2. Schematic diagram of a two-stage anaerobic digestion system for solid substrates.

sealed and the gas output and CH_4 content monitored. A temperature of 28-30°C was maintained throughout the digestion period by circulation of 1/3 of the liquid volume to an external heat exchanger every 3 days. Methane production was detected after 5-7 days and continued until day 58-60. The average methane content of the gas produced was 75% with a maximum of 88%. The volume of CH_4 produced in the single stage system was approximately 0.1 m^3/Kg T.S. compared with a yield of 0.16 m^3/Kg T.S. straw or 0.24 m^3/Kg T.S. grass by the two-stage system.

Full-scale application of the two-stage system for biomass digestion envisages the inclusion of a first-stage hydrolysis reactor alongside a full-scale fixed bed reactor whose primary function is the digestion of pig slurry in an intensive pig-rearing unit. Assessment of the operation of a full-scale fixed bed reactor is therefore an essential aspect of the project. A 20 m^3 reactor was designed and installed in a large pig-fattening unit. Raw slurry, after 2-3 weeks liquefaction in a holding tank, was passed to a slurry separator which removed the coarse fibre and hair fraction. The solids content of the liquid fraction obtained was higher than expected and averaged 4% T.S. Settleable solids were approximately 3%. The digester was operated initially on a 10-day liquid retention cycle and a gas output of 0.28 m^3 biogas/Kg T.S. or 0.43 m^3 biogas/Kg V.S. was obtained at an average loading rate of 5 Kg COD/m^3/day. Attempts to decrease the liquid retention time to 3 days resulted in a sizeable reduction in specific gas yield.

2. Conclusions

The data obtained to date suggest that phase separation of the hydrolysis/acidification and the methanation stages of digestion are feasible in the context of biogas production from solid biomass substrates. The optimum cycle of liquid displacement which will ensure maximum substrate liquefaction while simultaneously preventing the onset of methanogenesis needs to be determined for a variety of different substrates. Full-scale operation of the fixed bed reactor, although yielding results which are comparable to high-rate conventional systems, was complicated by the high solids content of the separated feed. Some design modifications are currently being assessed in an effort to maintain the specific gas yield at lower retention times.

References

1. Andrews, J.F. & Pearson, E.A. (1965) Int. J. Air Water Pollut., 9, 439.
2. Toerien, D.F. & Hattingh, W.H.J. (1969) Water Res., 3, 385.
3. Hobson, P.N. (1973) Process Biochem., 8, 19.
4. Lawrence, A.W. & McCarthy, P.J. (1969) J. Water Pollut. Control Fed., 41, 1.
5. Ghosh, S. & Klass, D.L. (1976) Paper presented at the 47th Annual Conference of the Water Pollution Control Federation, Denver, October 6-11, 1974.
6. Ghosh, S. & Klass, D.L. (1978) Process Biochem., 13, 15.
7. Newell, P.J., Colleran, E. & Dunican, L.K. (1979) Anaerobic Digestion Poster Papers, p.17, A.D. Scientific Press, Cardiff, Wales.
8. Colleran, E., Booth, A., Barry, M., Wilkie, A., Newell, P.J. & Dunican, L.K. (1981) Proceedings of Biomass, Brighton, 4-7 November, 1980. (In press).

METHANE PRODUCTION BY ANAEROBIC DIGESTION OF ALGAE

Authors : C.-M. Asinari di San Marzano, A. Legros, C. Piron, C. Sironval, E.-J. Nyns and H.P. Naveau

Contract Number : ESE-R-025-B (G)

Duration : 12 months 1 July 1980 - 30 June 1981

Total Budget : 8.500.000 FB CEC Contribution : 4.250.000 FB

Head of Project : Prof. E.-J. Nyns

Contractor : Catholic University of Louvain

Address : Unité de Génie Biologique
Place Croix du Sud, 1, bte 9,
B-1348 Louvain-la-Neuve, Belgium

Subcontractor : Prof. Sironval
University of Liège

Summary

The research aims (1) at improving yields and rates of biomethanation of algae and (2) at correlating composition of algal batches with methane productivities. Previously, it had been shown with *Hydrodictyon* algae that a two step system (liquefaction and biomethanation) gave improved results over a one step, completely-mixed system. In the present work, the influence of environmental factors such as temperature and load on the first step were investigated. Although the results of the liquefaction varied somewhat with the running conditions, these variations were without appreciable effect on the overall methane productivities. Besides, preliminary results of a two step system , a first percolating step followed by an upflow methane digester with active biomass accumulation look promising. Application of the two step biomethanation process to the algae *Tetraselmis* elicited a problem of reliability with time, presumably due to enhanced concentrations in ammonia resulting from increased biodegradation of proteins. Increasing the salt (NaCl) content (presently up to 20 g x l^{-1}) in the mixed liquor of a one step digester had no influence on the process. Biomethanation of six algal species showed good although inequal potential for methane production with yields from 0.20 to 0.33 l CH_4 x g^{-1} VS_0. Monitoring based on volatile solids (VS) contents appears erratic in the two step process because samples contain variable amounts of compounds volatile below 100 °C. Monitoring based on COD appears more reliable. Lagoons for the production of ton amounts of freshwater *Hydrodictyon* algae in industrial luke-warm were constructed. Production and population analyses of the algae are in progress.

1. INTRODUCTION

1.1. Past results. It has been shown (1) that both rate and yields of methane production from the freshwater algae *Hydrodictyon* can be increased by implementing a two step biomethanation system, each stage of which consists of a completely-mixed, semi-continuous fermentation without active biomass recycle. *E.g.*, the yield at 14 d mean retention time and C_V(*) = 4 g VS_0 x l_{ML}^{-1} x d^{-1} increases from 0.20 in the single step process to 0.33 l CH_4 x g^{-1} VS_0 in the two step process. Moreover, the reliability of the two step biomethanation system appears improved as compared to the one step, completely-mixed biomethanation system.

1.2. Objectives of the research. Five objectives are assigned to the present research : (1) optimisation of the two step biomethanation system (liquefaction and digestion), applied to the freshwater algae *Hydrodictyon*, (2) optimisation of the biomethanation of the marine algae *Tetraselmis*, (3) evaluation of the potential of other algal species, produced or harvested within the EC Programme, to be bioconverted into methane, (4) nitrogen and carbon mass balances around the biomethanation system for a better assesment of the process and (5) production of freshwater algae and study of variations in their composition.

2. MATERIALS AND METHODS

2.1. Algae. Unicellular marine algae *Tetraselmis* are grown in ponds at Lamezia Terme, (Calabria, Italy) (Prof. K. Wagener and R. Materassi, contract ESE/R/020/D). Once harvested they are oven-dried, air-dried or kept fresh after centrifugation. *Porphyridium cruentum* is cultivated in batch in artificial seawater and *Scenedesmus acutus* grown in continuous freshwater cultures (Dr. Gudin, contract ESE/R/037/F). They are harvested by centrifugation. The freshwater algae are filamentous *Hydrodictyon reticulatum* which grow in luke-warm water flowing through lagoons near the power plant of Tihange (Belgium), and originating from the river Meuse. The later algae are sometimes mixed with other algae such as *Cladophora*, diatoms and even higher plants like *Lemna*. The collected green biomass contains small animals, especially gasteropods which feed on *Hydrodictyon*.

2.2. Population analyses of *Hydrodictyon* algae. Randomised samples are taken each week either from the fresh or the dried biomass and are analysed for population and chemical analyses as follows. For population analyses, fresh 100 g samples collected at random are washed several times with tap water and filtered on paper. The species selected for the analyses (*Hydrodictyon, Cladophora,* other algae, *Lemna,* gasteropods and insects) are separated by hand and dried at 85 °C untill constant weight. Chemical analyses include total volatile matter, proteins, lipids and sugars, ashes and minerals. As a control, the water of the river Meuse is analysed for its mineral content.

2.3. Running conditions of methane digester. Completely-mixed, one step and second step methane digesters are operated anaerobically at 35 °C, with semi-continuous loadings and no active biomass recycle in 2 l all glass vessels described previously (1).

3. RESULTS

3.1. Optimisation of the two step biomethanation system using either *Hydrodictyon* or *Tetraselmis*

3.1.1. Optimisation of the first step (liquefaction) of the two step process

(*) Symbols and abbreviations : see Table VIII

Optimum liquefaction was tentatively defined by the maximum organic matter, determined as COD, released from the algae in a soluble form. A number of parameters involved in the definition of the best running conditions for an optimal liquefaction were investigated. Among these were the volumetric load, C_V, the mean retention time, θ, the temperature, T, as well as anoxy *vs.* anaerobiosis.

A first series of experiments was conducted in completely-mixed fermenters at room temperature under anoxic conditions with once a day loading and no active biomass recycle. The results of steady states are shown in Table I.

Table I. Results of the first step (liquefaction) of the two step biomethanation process using *Hydrodictyon* and *Tetraselmis* under various conditions

	C_V $\frac{g\ VS_o}{l_{ML} \times d}$	θ d	T °C	pH	COD_o $\frac{g\ O_2}{l}$	Y_{conv} $\frac{g\ COD_r}{g\ COD_o}$	Y_{liq} $\frac{g\ COD_s}{g\ COD_o}$	VA2	VA3 $\frac{g}{l}$
(1)	17.0	2	20	5.8	49.7	0.04	0.22	1.17	0.41
(1)	17.0	2	35	5.8	49.7	0.05	0.25	1.13	1.08
(1)(5)	11.3	3	55	5.8	49.7	0.10	0.32	1.62	0.35
(2)	7.5	4	20	5.3	42.5	0.07	0.16	2.37	1.40
(3)	13.0	2	20	6.8	38.1	0.08	0.20	1.87	0.79
(3)	28.0	2	20	7.0	74.8	0.04	0.20	3.46	1.21
(4)	28.0	2	20	7.0	83.0	0.17	0.31	4.27	1.87

(1) *Hydrodictyon*, anoxic liquefaction
(2) *Hydrodictyon*, anaerobic liquefaction
(3) *Tetraselmis*, air-dried, anoxic liquefaction
(4) *Tetraselmis*, fresh, anoxic liquefaction
(5) Still in progress at θ = 2 d and C_V = 17 g $VS_o \times l_{ML}^{-1} \times d^{-1}$

A second series of experiments was conducted in batch fermenters at room temperature. Suspensions containing per l about 25 g (VS) *Hydrodictyon* algae, of which respectively 12 % (triangles) and 56 % (circles) (wt. x $wt.^{-1}$) came from a previous 1st step mixed liquor were maintained either in anoxy (open symbols) or in anaerobiosis (filled symbols) for 8 days. The concentration in usefull end products : lactate and volatile fatty acids is given in Fig. 1. Owing to the fast production of volatile fatty acids at the onset of the liquefaction reaction and the later decrease in their concentration, a liquefaction system based on a percolation technique is being tried. It allows short mean retention times for the percolating fluid, distinct from longer solids mean retention times. Running conditions and preliminary results are given in Table II.

3.1.2. Optimisation of the overall 2 step process with *Hydrodictyon* algae.

Mixed liquor from the liquefaction step at 20°C were hitherto biomethanised in completely-mixed digesters (1). For the purpose of comparison, mixed liquors from liquefaction at 35 °C (Table I line 2) were biomethanised in the same conditions. Although the liquefaction yield as g $COD_s \times g^{-1}\ COD_o$ is somewhat improved by the higher temperature, the final methane yield does not change (Y_{ec} = 0,26 l $CH_4 \times g^{-1}\ VS_o$). Conversion and efficiency remain identical with however a somewhat lower concentration in volatile fatty acids in the effluent mixed liquor. Besides, experiments are being conducted using as second step an upflow methane digester with active biomass accumulation. The latter digester is fed either with the soluble portion of a completely-mixed liquefactor or the fluid from a percolating system (§ 3.1.1.). Results are given in Table II.

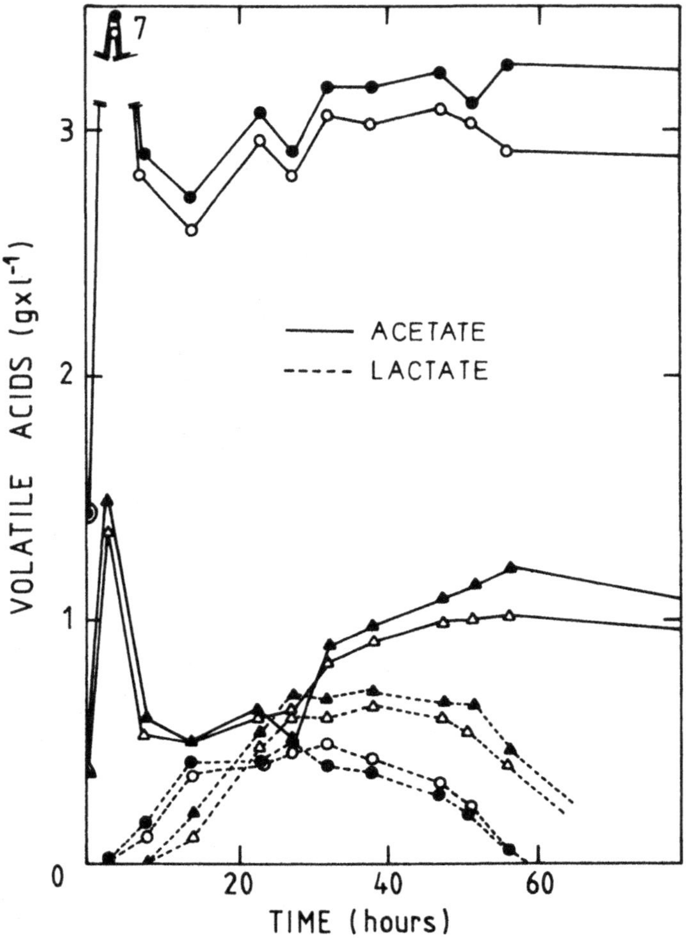

Fig. 1. Characterisation of first step (liquefaction) batches

3.1.3. Biomethanation of the marine algae *Tetraselmis*. As biomethanation of freshwater *Hydrodictyon* algae yields better performance in a two step syteme and that a first liquefaction step could anyway correspond to a storage between harvest and process, the comparison between one and two step biomethanations was extended to oven-dried, sun-dried and fresh *Tetraselmis* algae. The results are reported in Table III. Furthermore biomethanation in artificial seawater is in progress. Salt content is increased by steps of 5 g x l^{-1} over 2 weeks with 2 weeks of stationary state between 2 increases. Up to 20 g NaCl x l^{-1}, with C_V = 2 g VS_o x l_{ML}^{-1} x d^{-1} and θ = 14 d, no change in yield (Y_{ec} = 0.25 l CH_4 x g^{-1} VS_o) has as yet been observed.

3.2. Evaluation of the biomethanation potential of other algae. Up to now six algal species have been evaluated in biomethanation trials to check the potential of methane production of these energy crops, using completely-mixed, semi-continuous one step methane digesters at low loads. Results are given in Table IV.

Table II. Two step biomethanation of the freshwater algae *Hydrodictyon* using an upflow methane digester with active biomass accumulation, fed with the soluble portion from a completely-mixed liquefying fermenter or the fluid from a percolating liquefying fermenter

	Liquefying fermenter 1st step	Liquefying fermenter 2nd step	Percolating fermenter 1st step	Percolating fermenter 2nd step
C_V (g COD_o x l^{-1}_{ML} x d^{-1})	13.8	2.4		var.
θ_h (d)	4	5.7	0.33	1.25
θ_s (d)	4	25		
T (°C)	20	35	35	35
S_o (g x l^{-1})	55.1 (COD_o)		29.3 (VS_o)	
$Y_{liq.}$ (g COD_s x g^{-1} COD_o)	0.25			
VA2 + VA3 (g x l^{-1})	3.73	0.1	1.2	0
v_{CH4} (l CH_4 x l^{-1}_{ML} x d^{-1})		0.87		2.5-0.2
Y_{ec} global, (l CH_4 x g^{-1})	0.09 (COD_o)		0.22 (VS_o)	
[F_{420}] (µM)				0.24

Table III. Comparison between one and two step biomethanations of marine algae *Tetraselmis* in completely-mixed digesters. Liquefaction at 20 °C and biomethanation at 35 °C

	C_V	θ	F_{420}	v_{CH4}	Y_{ec}	Y_{conv}	VA2	VA3	N_{NH4^+}
	$\frac{g\ VS_o}{l_{ML} \times d}$	d	µM	$\frac{l\ CH_4}{l_{ML} \times d}$	$\frac{l\ CH_4}{g\ VS_o}$	$\frac{g\ VS_r}{g\ VS_o}$	$\frac{g}{l}$		
1. One step biomethanation									
(1)	1.0	20	(0.43)	0.21	0.21	0.27	0	0	
(2)	1.9	14	0.29	0.49	0.26	0.48	0.5	0	1.64
(3)	1.0	20	0.27	0.33	0.33	0.37	0	0	1.10
2. Two step biomethanation									
(2) 1st step	13	2		-	-	0.15	1.9	0.8	0.71
(2) 2nd step	1.	14	0.27	0.41	0.21	0.45	1.2	0.3	1.61
(2) 1st step	28	2		-	-	0.21	3.5	1.2	1.18
(2) 2nd step	4	14	0.65	0.77	0.19	0.36	2.7	1.0	2.29
(3) 1st step	28	2				0.29	4.3	1.8	1.57
(3) 2nd step	4	14	0.85	0.75	0.19	0.42	5.1	0.9	2.71

(1) : oven-dried; (2) : air-dried; (3) : fresh

3.3. Nitrogen and carbon mass balances. For a better assesment of yields and rates of biomethanation, accurate nitrogen and carbon mass balances appear essential. Kjeldahl nitrogen (N_K), ammonium nitrogen (N_{NH4^+}) and chemical oxygen demand (COD) determinations were made on effluent samples (out) from liquefaction fermenters and biomethanation digesters as well as on starting materials (in). Protein contents were estimated by multiplying the organic nitrogen (N_K - N_{NH4^+}) by the factor 6.25. Results of analyses for different types of digestion systems are shown in Table V.

Improvement of carbon mass balances were approached using COD and comparing the results with VS determinations. Results of conversion and liquefaction yields obtained by both methods for experiments reported in Table I and IV, are given Table VI.

Table IV. Biomethanation at 35 °C of various algal species in completely-mixed one step semi-continuous digesters without active biomass recyle

	C_V $\frac{g\ VS_o}{l_{ML} \times d}$	θ d	v_{CH4} $\frac{l\ CH_4}{l_{ML} \times d}$	y_{ec} $\frac{l\ CH_4}{g\ VS_o}$	Y_{conv} $\frac{g\ VS_r}{g\ VS_o}$	Y_{eff} $\frac{l\ CH_4}{g\ VS_r}$
Freshwater algae						
Scenedesmus acutus	1.0	20	0.20	0.20	0.45	0.44
Hydrodictyon reticulatum	0.9	20	0.19	0.21	0.61	0.35
Anabaena cylindrica	1.8	20	0.47	0.26	0.40	0.64
Marine algae						
Porphyridium cruentum	1.0	20	0.22	0.22	0.37	0.59
Laminaria saccharina	1.0	20	0.25	0.25	0.72	0.35
Tetraselmis (air-dried)	1.9	14	0.49	0.26	0.48	0.55
(fresh)	1.0	20	0.33	0.33	0.37	0.89

Table V. Nitrogen and protein mass balances in various liquefaction and biomethanation conditions

	C_V $\frac{g\ VS_o}{l_{ML} \times d}$	θ d	N_K in	N_{NH4^+} in	N_K out	N_{NH4^+} out	Proteins in	Proteins out	$\frac{in-out}{in}$ %
			$g \times l^{-1}$						
1. *Hydrodictyon,* completely-mixed biomethanation at 35 °C									
	1.9	14	1.36	0.04	1.50	0.72	8.2	4.9	40
2. *Hydrodictyon*, continuous liquefaction at 20 °C except lines 4 and 5									
Anoxic	28	2	4.80	0.39	3.70	0.64	27.6	19.1	31
Anoxic	10	4	2.25	0.14	2.56	0.53	13.0	12.7	2
Anoxic	17	2	2.37	0.06	2.28	0.44	14.4	11.5	20
Anoxic, 35 °C	17	2	2.37	0.06	2.38	0.40	14.4	12.4	14
Anoxic, 35 °C	17	2	2.37	0.06	2.23	0.43	14.4	11.2	22
3. *Tetraselmis*, one and two step biomethanation systems									
One step, 35 °C	1.86	14	2.66	0.04	2.70	1.64	16.4	6.6	60
Two steps									
1st, 20 °C	13.0	2	2.66	0.04	2.59	0.71	16.4	11.7	28
2nd, 35 °C	(1.86)	14	2.59	0.71	2.60	1.61	12.7	6.2	46
Overall			2.66	0.04	2.60	1.61	16.4	6.2	62

Table VI. Comparison between yields calculated from either COD or VS for various liquefaction or biomethanation conditions

θ d	S_o g VS_o	T °C	Y_{conv} $\frac{g\ COD_r}{g\ COD_o}$	Y_{conv} $\frac{g\ VS_r}{g\ VS_o}$	Y_{liq} $\frac{g\ COD_s}{g\ COD_o}$	Y_{liq} $\frac{g\ VS_s}{g\ VS_o}$	VA2 $\frac{g}{l}$	VA3 $\frac{g}{l}$
1. *Hydrodictyon*, liquefaction								
4	40	20	0.04	0.25	0.25	0.16	2.50	1.23
2	34	20	0.04	0.12	0.22	0.25	1.17	0.41
2	34	35	0.05	0.19	0.25	0.20	1.13	1.08
3	34	55	0.10	0.26	0.32	0.26	1.62	0.35
2. *Tetraselmis*, liquefaction								
2	26	20	0.08	0.15	0.20	0.16	1.87	0.79
2	56	20	0.04	0.21	0.20	0.16	3.46	1.21
3. *Tetraselmis*, completely-mixed methane digestion								
14	26	35	0.56	0.48	-	-	0.50	0

Table VII. Composition(4) of the biomass collected at Tihange during the winter 1980-1981

Dates of harvest	22.09.80(1)	30.09.80(2)	06.10.80(3)
Hydrodictyon	86.5	80.0	53.8
Lemna	5.5	0.5	0.4
Gasteropods	7.5	13.3	22.5
Insects (larva)	0.5	5.2	20.9
Higher plants	0	0.6	2.0
Others algal species	0	0.6	0.3

Harvested at the surface and mean of 4 samples (1), as a mixture at the surface and the bottom and mean of 5 samples (2) and at the bottom and mean of 6 samples (3). (4) : dry weight as % of dry weight of total harvest. Temperature of the water in the lagoons : 15-17 °C.

Table VIII. Abbreviations and symbols

1. Concentrations

- VS : Volatile solid ($g \times l^{-1}$)
- VA2 : Acetic acid($g \times l^{-1}$)
- VA3 : Propionic acid($g \times l^{-1}$)
- F_{420} : Coenzyme F_{420} (μM)
- S_o : Load ($g\ VS_o \times l^{-1}$)
- COD : Chemical oxygen demand ($g \times l^{-1}$)
- N_K : Kjeldahl nitrogen ($g \times l^{-1}$)
- N_{NH4^+}: Ammoniacal nitrogen ($g \times l^{-1}$)

2. Mean retention time

- θ : Mean retention time, h = hydraulic, s = solids (d)

3. Rates

- C_V : Volumetric load ($g\ VS_o \times l_{ML}^{-1} \times d^{-1}$ or $g\ COD_o \times l_{ML}^{-1} \times d^{-1}$)
- v_{CH4} : Methane production rate ($l\ CH_4 \times l_{ML}^{-1} \times d^{-1}$)

4. Yields

- Y_{liq} : Solubilisation yield ($g\ COD_s \times g^{-1}\ COD_o$)
- Y_{ec} : Yield ($l\ CH_4 \times g^{-1}\ VS_o$ or $l\ CH_4 \times g^{-1}\ COD_o$)
- Y_{conv}: Conversion ($g\ COD_r \times g^{-1}\ COD_o$ or $g\ VS_r \times g^{-1}\ VS_o$)
- Y_{eff} : Efficiency ($l\ CH_4 \times g^{-1}\ VS_r$)

5. Indices

- o : Influent
- r : Removed
- ML : Mixed liquor
- s : Soluble portion

3.4. Production and population analyses of the freshwater algae *Hydrodictyon reticulatum*. Fifteen lagoons were constructed at Tihange (Belgium). They are 15 to 40 m long, 5 m large and 0.5 m deep and cover about 2 500 m^2. The water, arriving from the cooling circuit of the power plant, enters the lagoons at a temperature which is 12 °C higher than that of the river Meuse from which it is taken. The water flow is 50 $l \times min^{-1}$ (3.3 $l \times min^{-1} \times lagoon^{-1}$). In June 1981, the temperature of the water was 25-28 °C and the algae completely filled the surface of the lagoons. The first harvest occured on June 15th. The production schedule includes biomass harvests at least once a week from June to September included. Harvest is followed by gentle centrifugation (in order to eliminate the bulk of the water from the fresh biomass) and by drying in the open air before storage. One expects to collect about 2 5 t of dried biomass in 1981, i.e.a yield of 10 to 12 $t \times ha^{-1} \times year^{-1}$.

The population composition of three samples taken during the winter 1980-1981 is presented in Table VII. The dry weight of the different categories is expressed as % of dry weight of the total collected biomass.

4. CONCLUSIONS

In the two step biomethanation of algae, the first step, liquefaction, is markedly influenced by increasing temperature. Both yield, expressed as the relative soluble COD released, and losses (as COD conversion), increase. A limited supply of air improves the liquefaction as yields obtained in anoxy appear higher than in anaerobiosis. Furthermore, there is no apparent connection between conversion, liquefaction yields and the productivity of volatile fatty acids (Table I).

Considering the complete two step system, the overall methane yield remains unaltered irrespective of the conditions of the liquefaction (Table II and § 3.2.). At high loads with *Tetraselmis*, the two step system appears less reliable than with *Hydrodictyon*. A tentative explanation is the higher concentrations of ammonia - and thus a toxicity effect - observed in the methane digestion mixed liquor in the former case and arising from an increased relative protein biodegradation (Table V, respectively 40 and 60 % with *Hydrodictyon* and *Tetraselmis*. A possible solution lies in an acclimation of the bacterial community.

The first trial of methane digestion of the soluble portion of the first liquefaction step in an upflow digester with biomass accumulation shows promising,even better with the fluid from a first percolating step, as indicated by the good economical yield (0.22 l CH_4 per g VS_0) and the absence of volatile fatty acid in the effluent methane digestion liquor.

Progressive acclimation of the bacterial community of a completely mixed one step methane digester to marine salt conditions has hitherto no adverse effect on the methane productivity, and the concentration of volatile fatty acids in the effluent mixed liquor remains negligible.

All algal species evaluated so far show a good although inequal potential for methane production.*E.g.*, although polysaccharides had been previously partially extracted from *Porphyridium* algae, their concentrations in the feedstock for biomethanation remains sufficiently high even at low volumetric load (1 g VS x l_{ML}^{-1} x d^{-1}) to produce a gelatinous mass which hampers manipulation and analyses. Therefore, the evaluation for potential methane productivity from *Porphyridium* is being resumed with a feedstock now completely extracted and hence containing lower amounts of carbodydrates.

Mass balances around a methane digestion system based on the determination of volatile solids (VS) are erratic because the samples to analyse do contain variable amounts of organic matter volatile below 100 °C such as lower fatty acids. On the contrary mass balances based on the determination of COD give more reliable results (Table VII). The latter method now replace routinely the former one.

Lagoons for growing *Hydrodictyon* (2 500 m^2) were constructed and are being put into use. An annual production of 2.5 t of dry biomass is expected. Population analyses made in winter show a variation which might be related to depth of sampling or time of collection. Characteristics of the composition of the biomass need to be correlated with parameters of the biomethanation.

5. REFERENCES

(1) Methane production by anaerobic digestion of algae, H.P. Naveau, C.-M. Asinari and E.-J. Nyns. C.E.C., Energy, 1st Coord. Meet. Contract. Energy from Biomass (Project E), Amsterdam, Sept. 18-19, 1980, Proc., Doc. EUR 7160 EN, DG XII, ed., 183-194 (1981).

TWO-PHASE PROCESS FOR THE ANAEROBIC DIGESTION OF ORGANIC WASTES YIELDING METHANE AND COMPOST

Contract Number	:	ES-E-R-040-N (N)
Duration	:	36 months 1 July 1980 - 30 June 1983
Total Budget	:	Fl. 820.000 CEC Contribution : Fl. 410.000
Head of Project	:	Drs. Berend A. Rijkens, IBVL
Contractor	:	Institute for Storage and Processing of Agricultural Produce (IBVL)
Address	:	Institute for Storage and Processing of Agricultural Produce (IBVL) Bornsesteeg 59 P.O. Box 18 6700 AA WAGENINGEN The Netherlands.

Summary

The aim of this project is to build and to operate a pilot plant in order to develop a good technical concept for medium and large scale operations, and to produce a feasibility study concerning this novel process. The laboratory work, including some semi-technical operations, was aimed at the degradation processes in the batch reactor R_1 and at the response of the "system" (the R_1 linked to the upflow sludge blanket (u.s.b.) reactor R_2) to various operation conditions. A series of experiments with various hydraulic retention times (h.r.t) did not show a significant influence on the rate of breakdown in the R_1, the addition of alkali does not seem to be necessary and in respect to the temperature range, mesophilic is definitely faster than thermophilic and more dependable. These last results will have to be confirmed, since they differ from indications we received before. The pilot plant with an initial capacity of 700 ton horticultural wastes per year is under construction and will be commissioned in August 1981. The R_1 is a cheap "clamp-type" reactor, with side walls of standard concrete elements, covered with a plastic sheet. The R_2 is a commercially available steel reactor of 10 m^3 volume. The preliminary feasibility study indicates that this process may be a good alternative for the processing of putrescible wastes.

1. The Experimental Work

1.1 Introduction

The novel two-phase process for the anaerobic digestion is tested in the laboratory with promising results. A pilot plant is under construction. The final goal of the project is to develop a good technical concept and a feasibility study for medium and large scale operations.

The principle of the process is that solid organic wastes, as delivered, are digested anaerobically in a batch reactor R_1 under a continuous percolation of water. The effluent of the R_1, containing dissolved organic matter, is treated in a conventional upflow sludge blanket (u.s.b.) reactor R_2 and recycled to the R_1. Initially biogas is developed in the R_2, but after a few days the R_1 also begins to produce biogas and eventually will become completely self-supporting.

1.2 Description of the apparatus and measuring equipment

The systems used in the laboratory (fig. 1) consists of two reactors, the R_1 (12 l.), the R_2 (20 l.) and ancillaries like a sandtrap, a peristaltic pump and an open gas seal maintaining a slight overpressure in the gas space of the R_1. In the tubing connecting the reactors to the Mariotte-bottle and the gas meter exchangeable gas sampling tubes are installed. The systems are placed in cupboard-like incubators (35°C or 60°C) or in a large walk-in incubator (35°C).

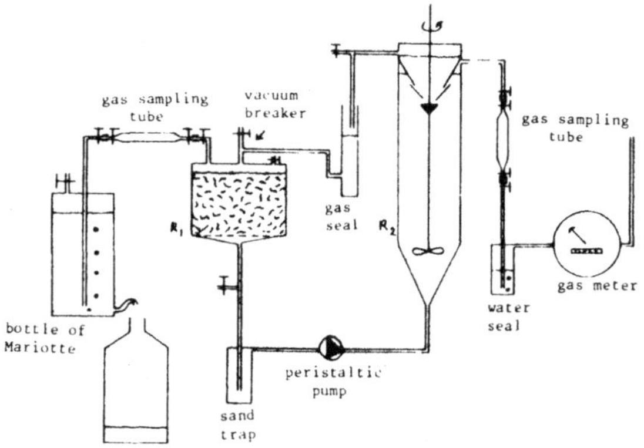

Figure 1. Schematic arrangement of the apparatus in the laboratory. The systems consists of the batch reactor R_1 (12 l.) and the continuous u.s.b. reactor R_2 (20 l.).

Semi-technical experiments are carried out in 200 l. steel vessels from an old uncommissioned pilot plant and in 350 l. vessels of glass-fibre reinforced resin.

The pilot plant under construction will have several clamp-type reactors R_1 and one R_2 (fig. 2). The first experiment will start with one R_1 (150 m^3 - 10 x 5 x 3.5 m) made of standard concrete elements with a drainage system at the bottom and spraying tubes resting on top of the waste. The reactor is covered with a plasticised sheet. The R_2 (10 m^3) is a commercial steel u.s.b. reactor. Ancillaries are two

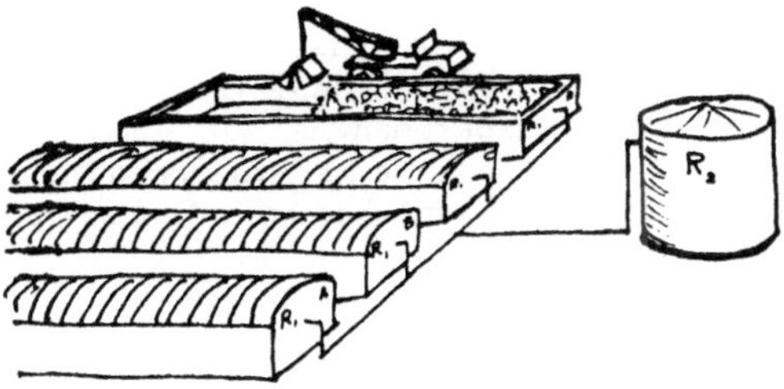

buffertanks (10 m^3 each), a heat exchanger, a boiler, a pumping station, etc. The plant will be commissioned in August 1981. Preliminary experiments are carried out in an adapted steel container of 50 m^3.

Figure 2. The concept for the pilot plant.

1.3 Results

The preliminary research carried out with various wastes(municipal solid waste - m.s.w., straw, sugar beet pulp, waste plant material, cannery waste, etc.) proved the need for a standard feedstock, available all year round. A standard compound "p.s.m.", consisting of dehydrated sugar beet pulp (50%), wheat straw (35%) and cow manure (15%), shows a reasonable similarity with various wastes (fig. 3).

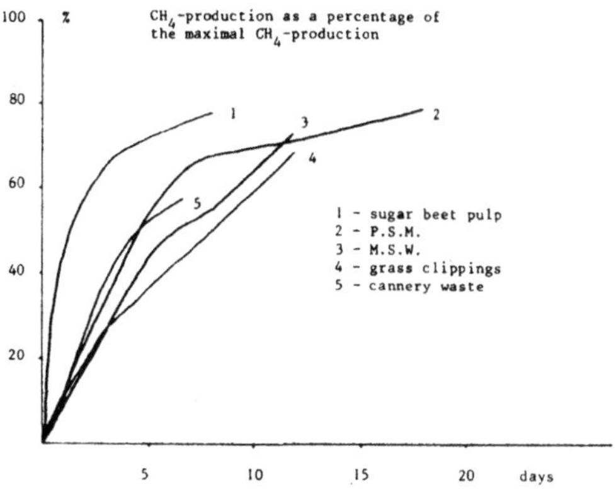

The COD is a useful parameter to indicate the mass flows in the system. For a better comparibility of experiments we introduced the "COD digestible" (COD_{dig}), being the amount of organic matter digestible in a reasonable time. Assumed is that all the components are digestible, except for the lignin and an equal weight of cellulose. The COD_{dig}, to be calculated from the analysis, generally amounts to 70 to 75% of the COD. The maximum CH_4 production can be calculated from the COD_{dig} (1 g. $COD_{dig} \equiv 0.375$ l. CH_4); the rate of breakdown can be expressed as a percentage of these values.

Figure 3. "Trend graphs" showing the digestion of various solid wastes.

The influence of the hydraulic retention time (h.r.t.) on the rate of the breakdown in the R_1 is tested in the range 8 to 33 h.; no significant effect was found, neither mesophilic nor thermophilic. Fig. 4 shows the "zones", indicating the scatter in a number of parallel experiments.

The influence of mesophilic and thermophilic conditions (35°C and 60°C) are also shown in fig. 4. Thermophilic appears to be slower and also more sensitive; a "damaged" flora recovers slowly and is often "total loss". These results are contrary to earlier findings, a check is necessary.

The addition of alkali, assumed to enhance the start of the methane flora, may not be necessary. Fig. 4 shows the rapid start of the methane production in the R_1, while fig. 5 indicates a reasonable rapid recovery from the lower pH at the start. However, this aspect will be investigated further. The "trend graphs" in fig. 5 depict the main characterising parameters from a series of experiments.

The drainage system for the pilot plant was tested in the 50 m^3 container filled with waste tomato plants. A sand bed with drain pipes proved to function adequately.

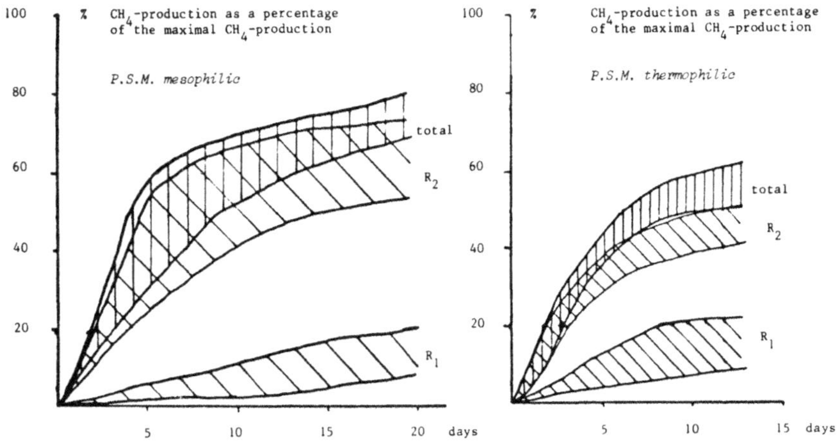

Figure 4. The digestion of p.s.m. at mesophilic and thermophilic temperature. The "zones" indicate the scatter in parallel experiments.

1.4 Analysis of results and comments

The breakdown of the solid matter in the R_1 starts with a high rate; 20 to 50% of the COD_{dig} is liquified, dissolved, carried away to the R_2 and converted to biogas in 2 to 7 days. In this period the readily degradable matter is leached, leaving a cellulose residue digesting at a much lower rate. Since anaerobic bacteria grow slowly the first attack can be attributed mainly to the seeded population, apparently very active. The population operating in the second phase may also be active, but since the degradation of "cellulose" basically is a slow process it might be rate determining.

The amount of organic matter, liquified in the R_1 equals the COD in the effluent plus the methane developed in the R_1. So the rate of decay in the R_1 is rather accurately depicted by the total methane production, assuming the R_2 performs well.

The nett costs for this process are estimated at about Dfl. 30 per ton solid waste. This price is somewaht higher than the dumping costs in the Netherlands today, but lower than controlled dumping and aerobic composting. The price for the anaerobic compost in the calculation is assumed to be nihil, the sale price is taken equal to the transportation costs.

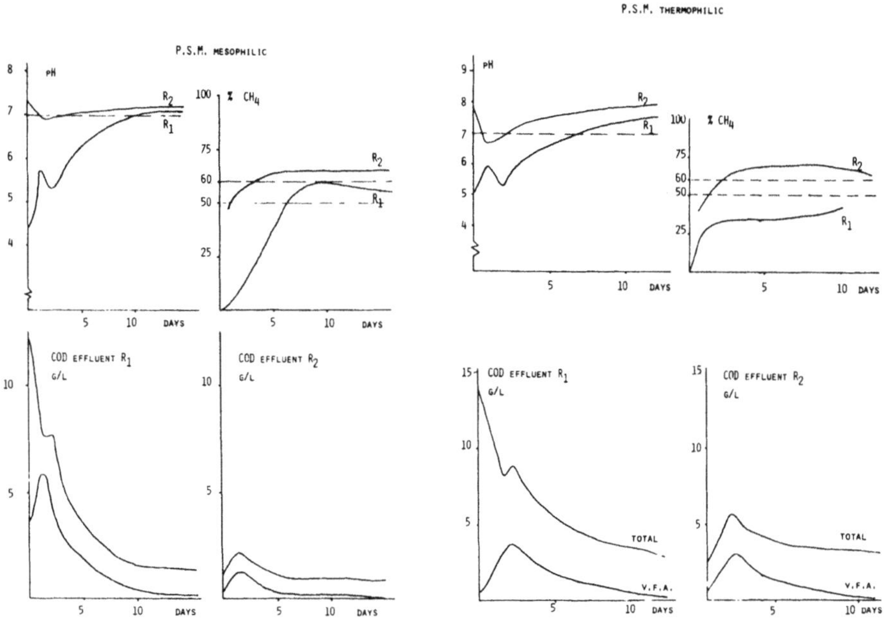

Figure 5. The digestion of p.s.m. at mesophilic and thermophilic temperature. "Trend graphs" of the main indicative parameters.

2. Conclusions

The breakdown of the readily degradable organic matter in the R_1 may result mainly from the seeded bacteria, hardly influenced by the percolation rate (h.r.t.) of the R_1.

In 7 days generally about 50% of the "COD digestible" is liquified, leached and converted to biogas; in 10 days about 70% is digested.

After this period the "cellulose" is attacked at a much lower rate in which the breakdown of the "cellulose" could be the determining step.

The rate of degradation in the R_1 is not influenced by h.r.t.'s in the range of 8 to 33 h. Alkalisation may not be helpful to speed up the digestion process.

The thermophilic process appears to work slower, is more sensitive and liable to disorder.

STARTING UP OF A THERMOPHILIC ANAEROBIC DIGESTION

Contract number : ESE- R - 039

Duration : 36 months, June 1980 - June 1983

Total budget : 180,000 units of account (50 % EC-contribution)

Head of project : Dr. G. Lettinga

Contractor : State Agricultural University

Address : Salverdaplein 10
6701 DB Wageningen

Summary

As a first item of our investigations concerning the feasibility of thermophilic anaerobic digestion for methane generation from aqueous wastes the start up of a thermophilic digestion was investigated at 55 °C. Three different seed materials proved to be essentially the same as a source of thermophilic methane bacteria. In batch experiments with acetic (C2), propionic (C3) and butyric (C4) acid as feed no simultaneous breakdown of C3 and C4 could be obtained, while in continuously fed mixed tank reactor (MTR) experiments at a high LRT (liquid retention time) and a loading of 80 mg/l.d of C2, C3 and C4, a complete breakdown of C3 was achieved within 33 days.
In addition an UASB-reactor was started up with mesophilic sludge, and it showed very similar patterns in the breakdown of the volatile fatty acids as obtained in the continuously fed MTR. The loading rate of the reactor can only be increased fairly slowly, which apparently conflicts with the high growth rates we measured. Possible reasons for this phenomenon are discussed.

1.1. Introduction

Thermophilic methane digestion, i.e. methane digestion in the range of 45 - 70 °C, may become an attractive alternative for mesophilic digestion in view of the high loading rates that can be achieved, the higher methane yields from complex wastes, compared to mesophilic digestion, the low number of pathogens in digested sludge and effluent and the small quantity of sludge produced. Moreover in various cases waste water temperatures are in the thermophilic range, or can be applied relatively simply by using available waste heat. Important drawbacks are the high costs for heating the influent in the case that waste heat is not available. Moreover the process seems to be more sensitive than the mesophilic digestion process.

The aim of the present investigations is:
(i) to assess a proper seed material to obtain a stable thermophilic digestion process
(ii) to develop a thermophilic sludge with good settling properties, as this is necessary in order to achieve high loading rates in anaerobic wastewater treatment systems, the UASB-concept included (1).

Emphasis was given to the acetate-producing and methanogenic steps of the digestion process, because these steps are usually rate-limiting in the treatment of chiefly dissolved wastes. This has been done by using a mixture of acetic, propionic and butyric acid as substrate.

For the sake of cenvenience, all experiments (except one) were carried out at 55 °C. This is a temperature well within the limits of the thermophilic growth range and it is supposed to be near the optimum from energetic as well as biological point of view (2,3)

1.2. Materials and methods

Most experiments were conducted in 6 l gas tight digesters with a net working volume of 5.5 l, which were stirred intermittantly (5 sec. at 140 r.p.m. every 5 min.). Experiments with upflow systems were carried out in 5.75 l columns, 0.80 m in height, equipped with a central ax with a flat blade and a gas-solids separator device.

Feed solutions consisted of C2, C3 and C4-acid in a 1:1:1 ratio (in weight) in tap water, to which was added (in mg/l): NH_4Cl, 137; NaH_2PO_4, 37; KCl, 25; $MgCl_2.6H_2O$, 8; $CaCl_2.2H_2O$, 9; yeast extract (Difco) 100; and 1 ml/l of a trace metal solution according to Zehnder et al.(4).

Three different seed materials were used:
(i) fresh cow manure, stored at 4 °C for a maximum of 7 days,
(ii) granular sludge from a UASB-reactor treating wastewater from a sugar factory, stored 3 - 30 months at 4 °C,
(iii) sludge from the sludge digester of a municipal treatment plant.
As there seem to be no benefits in raising the temperature in small steps to the temperature chosen (5,6), it was always raised in one step to the working temperature.

Analyses. The methane production was determined with a wet test gas meter in the experiments with upflow systems after passage through a 3 M NaOH-solution and a column with granular soda lime. In digester experiments methane production was measured with a liquid displacement system, with a 0.4 M NaOH-solution.

Concentrations of volatile fatty acids (VFA) were determined on a Becker 417 gas chromatograph fitted with a Chromosorb 101 (80-100 Mesh) column, and a flame ionization detector. Column temperature was 190 °C; N_2, saturated with formic acid, was used as the carrier gas.

COD was measured according to Standard Methods.

1.3. Results

Preliminary experiments concerned the nature of the seed material. Cow manure, granular sludge, digested sewage sludge (see Methods section) and a mixture of these three seed materials were used in a concentration of 0.5 %(w/v). Rapid methane generation started after 8 to 11 days in all cases, except for the digested sewage sludge where the production started a few days later, as shown in fig. 1. After 3 to 5 batch feedings of 600 mg/l C2, C3 and C4-acid each, a quite stable pattern of VFA-breakdown was reached (fig. 2). Doubling times of the bacteria growing on C2 and C3-acid were 18 and 45 hrs. respectively, as calculated from the methane generation rates in different batch experiments. In no case a simultaneous breakdown of C3 and C4-acid was observed !

As it was believed that the main reason for this phenomenon had to be attributed to the applied feeding pattern, a continuously fed system in a MTR was started up with granular sludge (4.2 g VSS/l), using 80 mg/l.d of C2, C3 and C4-acid as feed.Indeed now a complete C3-acid breakdown was achieved within a period of 33 days. Moreover the thermophilic population readily adapted to higher loading rates after this period (fig. 3). An attempt to operate a semi-continuously fed MTR with a 5.5 day retention time with 1 g/l of C2, C3 and C4-acid as substrate was not successfull. Even after 50 days of operation, C3-acid was only broken down for 50 %.

Start up of a thermophilic UASB-reactor was conducted by filling a 5.75 l reactor with 1.5 l of granular sludge (approx. 70 g VSS/l), which was stored for 2.5 years at 4 ^{o}C. 1 g/l of C2, C3 and C4-acid served as substrate. The results are quite characteristic ! After a first rapid drop of the methane production in the first 5 days, probably due to the loss of activity of the mesophilic population, the methane production increases from 5 to 60 % of the COD-input from day 12 to day 15. And over 90 % is reached after 33 days (fig. 4a). Conform the results of the continuously fed MTR, C3-acid breakdown was complete after 33 days, but it remains the first compound to build up in higher concentrations after applying higher loading rates, i.e. after an increase in COD-loading from 2.1 to 4.9 kg COD/m^{3}.d, it took at least 17 days to obtain C3-acid concentrations below 200 mg COD/l (fig 4b). After stabilization, the difference between filtrated and unfiltrated effluent COD averages 170 mg COD/l (fig. 4c).

1.4. Discussion

Apparently there are no big differences between the three applied seed materials. One might conclude that any material containing large amounts of mesophilic methane bacteria can be used as a seed material for thermophilic digestion. As with mesophilic digestion, propionic acid seems to be the first compound to build up in high concentrations, even after a stable population is obtained. This phenomenon cannot be attributed to the growth rates of the bacteria involved in the breakdown of propionate, as doubling times of propionate degraders, and acetate and hydrogen using methane bacteria are 44.5, 18 and 5.8 hrs. respectively (7). It seems likely that the rapid growth of butyrate-degrading (and hydrogen producing) bacteria is responsible for this. With a higher rate of C4-acid breakdown the hydrogen pressure can become thermodynamically unfavourable for the C3-breakdown (8). Apparently the hydrogen consuming bacteria adapt slowly to a higher hydrogen evolution rate, giving rise to a more or less severe retardation in the growth of propionate degrading bacteria.

A rapid decay of the same bacteria, as generally observed under thermophilic conditions, could explain the fact that under the conditions applied no simultaneous breakdown of C3 and C4-acid occurred in batch experiments. The failure of operation of a semi-continuously fed MTR may be attributed to (i) retardation of C3-acid breakdown during rapid hydrogen evolution in the first hours after feeding and (ii) the extreme sensitivity of the C3-acid bacteria to oxygen (9) of which always small amounts entered the reactor during addition of the substrate.

2. Conclusions

Any material exerting a reasonable methane production in the mesophilic temperature range can be used as a seed material for thermophilic methane digestion. For the start of an UASB-reactor a sludge with good settling properties is preferable.

A stable thermophilic population can be cultivated within a period of about 5 weeks. Thereafter, increases in the loading rate will lead to a rapid response, except for the breakdown of propionate, which shows retarded responses at large increases, presumably due to too high hydrogen evolution rates as a result of butyrate breakdown.

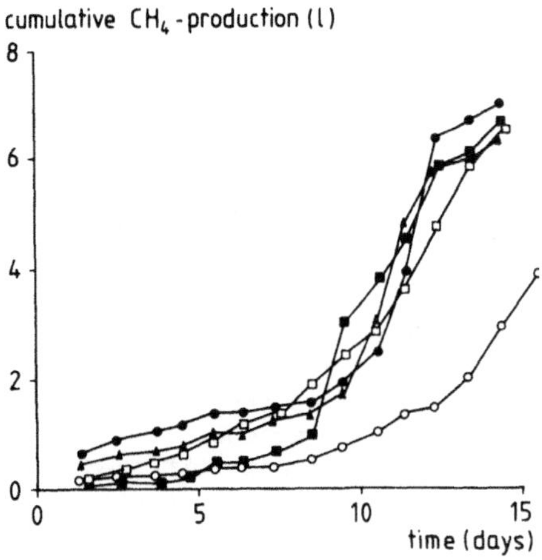

Fig. 1. Cumulative methane production as a function of time for different seed materials: cow manure (■), granular sludge (●), digested sewage sludge (○), mixture (▲), all at 55 °C, and cow manure at 60 °C (■)

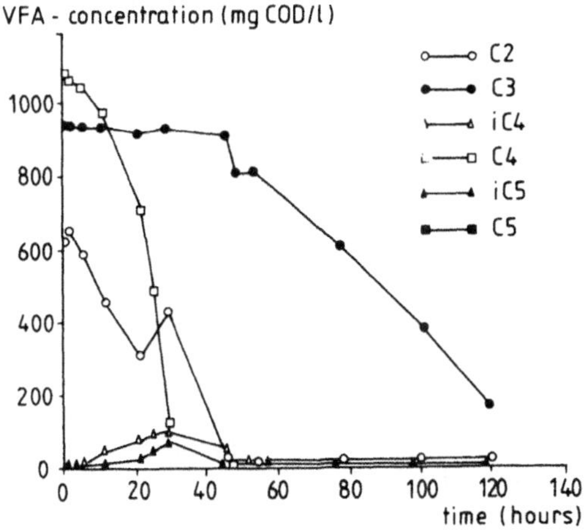

Fig. 2. Course of VFA-degradation after batch feeding of 600 mg/l of C2, C3 and C4-acid, after 47 days of operation with batch feeding only.

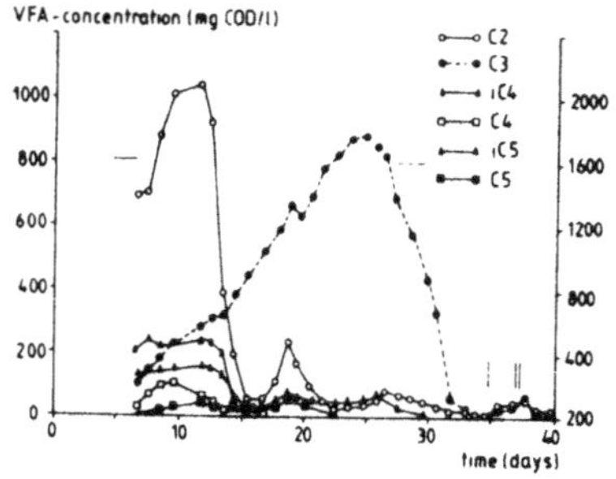

Fig. 3. VFA-degradation in a MTR with granular sludge (4.2 g/l). Feeding rate was 80 mg/l.d of C2, C3 and C4-acid, at a RT of over 100 days. At (↓) input was raised to 160 mg/l.d of each acid; at (↓↓) to 240 mg/l.d.

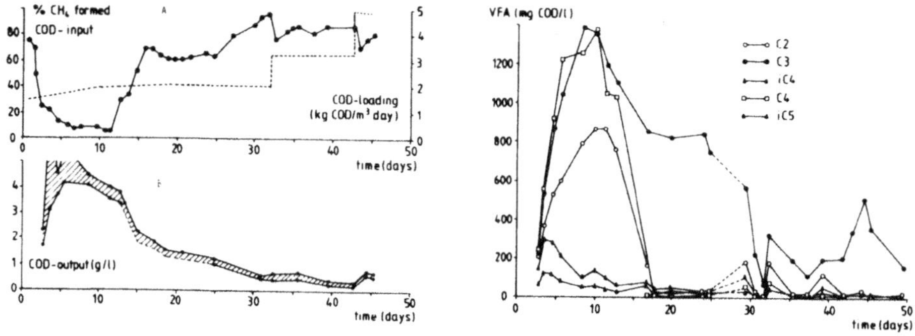

Fig. 4. Course of various parameters during start up of an UASB-reactor with mesophilic granular sludge, fed with 1 g/l C2, C3 and C4-acid. A: Methane production expressed as percentage of the COD-input (—); and COD-loading (----). B: VFA-concentrations expressed as mg COD/l in the effluent. C: COD of effluent, unfiltrated (upper line) and filtrated (lower line). The difference is a measure for the amount of bacteria in the effluent (shaded).

References

1. Lettinga,G. et al.(1980). Biotechn. & Bioeng. 22:699-734.
2. Varel,V.R. et al. (1980). Appl.& Environ. Microbiol. 40:217-222.
3. Pfeffer,J.T. (1974). Biotechn. & Bioeng. 16:771-787.
4. Zehnder,A.J.B. & K.Wuhrmann (1977). Arch. Microbiol. 111:199-205.
5. Garber,W.F. et al.(1975). J.Wat.Poll.Contr.Fed. 47:950-961.
6. Lettinga,G. (1980). Rep.Contractors Meeting Sept. 1980.
7. Zeikus,J.G.& R.S.Wolfe.(1972). J.Bacteriol. 109:707-713.
8. Kaspar,H.F. & K.Wuhrmann (1978). Microb. Ecology 4:241-248.
9. Boone,D.R. et al (1980). Appl.& Environ. Microbiol. 40:626-632

ALCOHOLIC FERMENTATION

IMPROVEMENT OF THE TECHNOLOGY BASED ON PHYSIOLOGICAL PHENOMENA

Contract Number : ESE-R-044-F

Duration : 24 months 1.07.80 - 30.06.82

Total budget : participation CEC 42 700 ECU

Head of Project : Dr. G. GOMA and J.M. NAVARRO

Contractor : Department Génie Biochimique Dr. Prof. G. DURAND
Institut National des Sciences Appliquées

Address : avenue de Rangeuil
F - 31077 TOULOUSE CEDEX

Summary

A major constraint of conventional alcohol fermentation processes is ethanol or end product inhibition and substrate inhibition. Our work is developped at the level of 1) a better comprehension of mechanisms of inhibition(intracellular ethanol accumulation related to environmental conditions : temperature substrate and kind of microorganisms) 2) increasing microorganism concentration by studies on microbial immobilization onto supports, 3) searching optimal technology of continuous bioreactors : combination of infinitely mixed reactors and plug flow reactors with divided feed and temperature profile, 4) developping technology of extrative fermentation, 5) improving cellulose utilization by clostridium thermocellum. Results obtained are descussed.

Ethanol technology will be improved if :

- Yields of fermentation are improved : 55 to 65% of ethanol cost is related to the raw materials
- Ethanol tolerance is improved (new strains, minimization of inhibitories effects) in the purpose of decreasing the cost of extraction and depollution
- Productivity is increased (decreases of volume of reactors and maintenance cost).

A multidisciplinary approach is necessary both at the technological level and physiological level.

1- Observations on the inhibition of alcoholic fermentation by ethanol

During fermentation, yeast growth is rapidly stopped when the concentration of alcohol in the medium increases but fermentive activity is not entirely inhibited until high alcohol concentrations are reached.

1-1 Intracellular ethanol : effect of temperature (1)(2)(3)

Several authors have shown that inhibitant action of ethanol is more important when temperature is hight and that alcohol added to culture medium is inhibiting less than ethanol produced by the cells during fermentation. So, the rate of alcohol accumulation within the cells and certain kinetic parameters were simultaneously determined in alcoholic fermentation processes using Saccharomyces uvarum cells. The growth inhibitory effect of alcohol was related to its retention inside within the cells, ie : yeast multiplication is stopped when intracellular alcohol reaches a maximum value. Moreover, the higher the temperature, the deeper the inhibitory effect of ethanol and the higher the maximal intracellular alcohol concentration.(Table 1). Activation energy determination showed that ethanol accumulation within the cells was a consequence of the resistance to its diffusion through the cell wall from within to outside the cells. Inorder to reduce diffusion resistance we added surfactants to the medium culture. Then yeast growth is not so rapidly stopped, ethanol production rate is increased and therefore fermentation time is shorter than is the reference assay(without surfactant ; temperature optimal)

Table 1 : Influence of the culture conditions on some parameters value of alcoholic fermentation by Saccharomyces uvarum

Culture conditions	Alcohol concentration (g/l)		Maximal biomass concentration g/l	(h) time without cell lysis	mean rate of cell lysis (g/l.h)
	in the medium	in the cells			
Temps 30°C	13	300	2,1	88	$38,2\ 10^{-4}$
Temps 20°C	25	170	2,6	190	$18,0\ 10^{-4}$
Temps 10°C	35	100	3,0	836	$5,2\ 10^{-4}$
Temps 30°C+ surfactant	48	120	4,5	0	$260\ 10^{-4}$

1-2 Effect of initial substrate concentration (4)(5)(6)

As an increase of temperature increases the intracellular accumulation, we have observed that when the initial sugar concentration (So) change maximal intracellular (Pim) ethanol changes. We study this phenomenon on 2 yeasts Saccharomyces cerevisiae (Sc) and Saccharomyces bayanus (Sb). The figure indicates the So effect on Pim. A maximum is obtained where So = 50 g/l. Saccharomyces bayanus which is more tolerant to ethanol accumulates less ethanol.

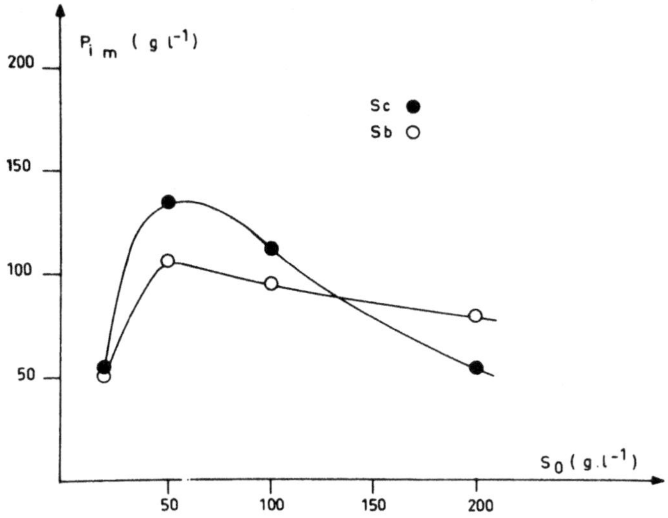

Figure 1

Change of the maximal ethanol accumulation intracell (Pim) with changes of initial substrat concentration (So).

We think that a part of substrate inhibition is correlated to the ethanol accumulate inside the cells. Initial substrate concentration has an influence on lipid composition of cell membrane.

1-3 Comments on the inhibition by added ethanol (7)

A lot of studies on ethanol inhibition are performed by addition of ethanol to the fermentation broth. This method is erroneous and we demondstrated a dual influence of added and produced ethanol. In our case, using Saccharomyces cerevisiae UG5 the growth rate (μ) was described by an hyperbolic relationship between μ and the ethanol concentration P.

$$\mu = \mu_m \frac{K_{ip}}{K_{ip} + P}$$

The calculated K_{ip} value related to ethanol added was 105 g/l (extrapolation) while the K_{ip} value to the produced ethanol was 3,5 g/l. The critical concentration of ethanol added is 85 g/l (concentration which stops growth and ethanol production) that means that the permeability of ethanol from outside the cells to the inside is very low.

We have to note the difference of sensibility of the cell to the extracellular and intracellular ethanol. The true sensibility of the strain to ethanol appears as dependent on the tolerance of the extracellular ethanol. Maximal ethanol concentration inside the cell is always greater than critical concentration of added ethanol. One explanation is that intracellular effect of ethanol inhibition occurs at the start of fermentation and during the first steps of fermentation while at the end of fermentation this phenomenon is less important and negligible ; so, as ethanol diffusion from inside to outside the cell is rate limiting, this limitation is enhanced by the outside ethanol. We are now investigating for several yeasts with different tolerance to ethanol the maximal internal ethanol concentration (P_{im}) and the critical external ethanol concentration (P_{ce}) ; ethanol tolerant cells have high P_{ce} and low P_{im}. If we compare Navarro's results on hexokinase inhibition by ethanol and these results we can understand the difference in cell growth inhibition and fermentation which is less sensiti ve to ethanol. The lethal effect of ethanol observed seems due not to the intracellular ethanol but to external ethanol by some "destructurant" effect of ethanol on cell wall and cell membran. Values of critical values obtained on several strains with differents tolerance to ethanol are given in table 2.

Table 2 : Critical values of ethanol concentration for differents strains.

Parameters / Strains	Maximal intracellular concentration observed (g/l)	Ethanol added with stop the growth (Pce g/l)	Maximal concentrations tolerated by the yeast P_p (g/l)
Saccharomyces sake	134	120	160
Saccharomyces bayanus	106	88	112
Saccharomyces cerevisiae UG5	132	86	96
Saccharomyces uvarum	300	45	60
Kloekera apiculata	-	20	32

Yeast which are more tolerant to ethanol are yeasts which accept high external concentration. There is corelation between ethanol tolerance and intracellular ethanol accumulation. Growth is first inhibited.

2- Improvement of cellulolysis by microorganisms (8)(9)(10)

Addition of surfactant Tween 80, Dimethyl acetamide (DMA), Dimethyl sulfoxyde (DMSO) improves the cell growth rate, the rate and amount of cellulase production of both Trichoderma reesei QH9414 and Clostridium thermocellum H7 grown on cellulose optimal amounts of DMA and DMSO are respectively 0,4 g/l and 0,2 g/l for both enhancement of the growth of Clostridium thermocellum and we discuss more the results obtained. First this strain growth easily only delignified lignocellulosic materials.

We compare results with sugar cane bagasse delignified or not and with cellulosic Whatman CC41 Table 3.

Table 3 : Yield in product (Ethanol + acid) for differents cellulosic materials. (θ : 60°C, pH 7,5, cellulose 10 g/l) Clostridium thermocellum M 7

	Products g/l	Percent of degraded substrate φ %	Yield g/g
Raw bagasse	1,05	20	0,5
Delignified * bagasse	3,09	69	0,41
Cellulose Wathamn CC41	3,18	82	0,32

* no growth if there is no washing.
The effect of additives is descrided by table 4.

Table 4 : Effect fo additives on final products of fermentation by Clostridium thermocellum M7 (Same conditions than for table 3 but So = 20 g/l).

	Acetic acid	Ethanol	Reducing sugar	ε P	φ	R
Reference	2	2	0,15	4,15	65	0,35
DMA 0,4 g/l	2,1	1,5	1,60	5,22	64	43
DMA 0,4 g/l Ergosterol 0,7 mg/l	2,0	1,82	1,50	5,32	60	44

Co fermentations of cellulolytic strains andalcohagenic strains are under studies.

3- Production of ethanol by immobilized cells (11)(12)(13)

In the usual manner , metabolic production is proportional to biomass quantity. Now, during alcoholic-fermentation, ethanol is inhibitor from growth and its production. So, this fermentation can be considered as a process having a very low productivity. In order to increase the biomass quantity inside the reactor we have immobilized yeast cells onto solid carriers as brick, wood chips, PVC....

Using two reactor types high concentrated ethanol solutions have been produced by Saccharomyces cerevisiae cells adsorbed outo wood ships. A column, where CO_2 produced by the reaction occupy a very important volume. More, a very bad distribution of the cells, which are produced in the lower part of the colum, when inhibition by the product is not important (solutions containing 13,5 % (V/V) of alcohol with bound the alcohol production to this part of the reactor. The best ethanol production has been obtained with an initial glucose concentration equal to 225 g/l and a productivity of 2,2 g of ethanol by hour and by liter of empty reactor.

In a multistage reactor (6 stirred tanks in series) CO_2 retention is limited and biomass distribution is better, so solution containing more than 17,5% (V/V) of ethanol have been obtained and then the productivity was 4,8 g/l.h.

4- Extractive fermentation (14)(15)

It is well known than ethanol is extractible from water by liquid-liquid extraction. Alcohol dissolves readily in some liquids that do not mix with water. By exploiting this solubility difference, alcohol can be recovered by solvent extraction, but for extractive fermentation the non toxicity of solvent toward the cells must also be established. After a screening work for research of compatible solvent we find dodecanol.

4-1 Choice of a technology of extractive fermentation

The technology of extractive fermentation coupling both liquid-liquid extraction and fermentation must have : (a) compatibility with a continuous process (b) plug flow hydrodynamics for the extractor liquid medium and the dodecanol.

We developed for SCP production a new type of continuous fermentor which features an agitation-aeration system hand on pulsed flow across perforated plates. This device was packed with porous brick in order to increase the biomass concentration in the reactor and to improve the liquid-liquid extraction by dodecanol. It is fed both by dodecanol and by fermentation medium. The device used has been run continuously during 1/2 years.

4-2 Improvement of alcoholic fermentation by extractive fermentation

Continuous cultures were performed with immobilized cells with or without extractive fermentation The results were obtained in extreme conditions for Saccharomyces cerevisiae.

Three series of experiments are reported, one of reference without dodecanol, the other, with only a tiny amount of dodecanol, for the purpose of evaluating the additive effect of dodecanol without any removal of ethanol, (surfactant effect of dodecanol, increase of the yeast clumping). The third series of experiments are extractive fermentation.

Results obtained are summarized in Table 5 ; obviously, alcoholic fermentation is improved by extractive fermentation. The residual sugar drops from 162 gl^{-1} to 2 gl^{-1} while the productivity is multiplied by 4 with input glucose at 260 gl^{-1} and by 5 at 317 gl^{-1}.

Table 5 : Continuous alcoholic fermentation with immobilized cells without dodecanol addition (control), with very small addition of dodecanol (experiment 1), with extractive fermentation (experiment 2,

3, 4). Yeast Saccharomyces cerevisiae UG5 mineral medium + 1 g/l yeast extract, θ 35°C, pH non regulated (3.2).

Experiments / Parameters	Control	1	2	3	Control	4
Dodecanol feed ($1h^{-1}$)	0	0,0285	1	2,55	0	1
Medium feed ($1h^{-1}$)	0,069	0,0667	0,0728	0,0698	0,068	0,068
Glucose input (gl^{-1})	260	263	260	263	315	317
Glucose output (gl^{-1})	162	165,9	2	1,8	242	2,9
Ethanol conc. in aqueous phase (gl^{-1})	38,1	39,8	21,2	9,4	28,9	23,7
Ethanol conc. in dodecanol phase (gl^{-1})		14,9	8,38	3,35		9,1
Total ethanol productivity (gl^{-1})	0,24	0,28	0,91	0,83	0,18	0,58
Ethanol yield constant ($g.g.^{-1}$)	0,4	0,47	0,51	0,5	0,4	0,5

5- Orientation of the works

Our actual work is now oriented on :

- coculture of cellulolytic or inulolytic and alcohogenic strains
- physiology of ethanol inhibition relation between cell membrann composition and ethanol diffusivity and tolerance
- mechanism of ethanol excretion from the cell
- optimisation by control computer of the continuous alcoholic fermentation with 6 reactors in series : effect of gradient temperature and substrate feed
- improvement of extractive fermentations.

Bibliographie

1- NAVARRO J.M., 1980, Fermentation alcoolique : influence des conditions de culture sur l'inhibition par l'éthanol. Cellular and Molecular Biology, 26, 241-246

2- NAVARRO J.M., FINCK J.D., 1981, Evolution de l'activité Hexokinase de "Saccharomyces uvarum" fermentant le saccharose. Soumis a cellular and molecular biology

3- NAVARRO J.M., 1980, Levures immobilisées : croissance et métabolisme ; fermentation alcoolique. Thèse Doct. es Sciences, Université Toulouse

4- MORENO M., Thèse Dr. Ingénieur, I.N.S.A. Toulouse n° , 1979

5- STREHAIANO P., G. GOMA, Proceding Colloque Société Française de Microbiologie Industrielle, Reims, 1981
6- STREHAIANO P., G. GOMA, submitted to American J. of oenology
7- NOVAK M., STREHAIANO P., MORENO M., GOMA G., Biotechnol. Bioeng. 23, 201-211, 1981
8- MAUGERI F., Thèse Dct. Ing. I.N.S.A. Toulouse, 1980
9- AL ANI D., Thèse Etat, U.P.S., Toulouse, 1981
10- AL ANI D., MAUGERI F., GOMA G., Coll. Cellul. Marseille, Mai 1980
11- RYU Y.W ., 1980, Production d'éthanol en continu à l'aide de levures adsorbées sur support solide. Thèse Doct. Ing. I.N.S.A. Toulouse
12- HARMANT P., 1980, Etude comparée de la fermentation alcoolique avec des levures libres et adorbées Thèses Doct. Ing. I.N.S.A. Toulouse
13- RYU Y.W., NAVARRO J.M., 1981, Etude comparée d'éthanol pour *Saccharomyces cerevisiae* immobilisé dans un réacteur tubulaire et dans un gradostat. S.M.I. Colloque Reims
14- MINIER M., Thèse Doct. Ing. I.N.S.A. Toulouse, 1981
15- MINIER M., G. GOMA, Under Press Biotechnol. letters.

UTILIZATION OF AGRICULTURAL WASTES AS ENERGY SOURCES

ENZYMATIC HYDROLYSIS OF CELLULOSE TO GLUCOSE

Authors : SECTION I : F. ALFANI, M. CANTARELLA, V. SCARDI
SECTION II : L. GIANFREDA, G. GRECO jr.

Contract number: ESE-R-041-I

Duration: 36 months 1 July 1980 - 30 june 1983

Total budget: 306 000 000 lira CEC Contribution: 100 000 000 lira

Heads of project: prof. Francesco Alfani
prof. Liliana Gianfreda

Contractor: Università degli Studi di Napoli

Address: Istituto di Principi di Ingegneria Chimica
p.le V. Tecchio
I-80125 Naples
ITALY

Summary

The research project deals with the enzymatic hydrolysis of cellulose to glucose in ultrafiltration membrane reactors. The reaction is catalyzed by a multienzymatic complex (endoglucanases, exoglucanases and β-glucosidase). Most cellulase complexes are deficient in β-glucosidase; furthermore, the latter enzyme is somewhat labile as regards thermal stability. Therefore, a two step process has been studied in which cellobiose conversion to glucose by β-glucosidase is performed in a membrane reactor which could make use of a stabilized enzyme. In a first reactor cellulose is converted to soluble fractions by native cellulase complexes.

The optimal operating conditions for both reactors have been separately studied (in terms of thermal stability of the enzymes, optimal pH, effects of product inhibition and optimal reaction temperature).

As far as the first reactor is concerned, the effect of several chemical and physical pretreatments on the rate of saccharification has been investigated. For the final reaction stage, the possible stabilization of β-glucosidase by soluble cellulose has been studied together with the effect of high molecular weight substrates accumulation at the membrane surface.

SECTION 1 -

EXTRACTION AND ENZYMATIC HYDROLYSIS OF NATIVE MIXED CELLULOSIC MATERIALS

1.1 Introduction

Most agricultural cellulosic wastes contain three major components, cellulose, hemicellulose and lignin in the ratios of roughly 4:3:3 (1), therefore they are interesting from an energetic point of view. In order to convert the cellulose into glucose and then, via fermentative routes, into ethanol or single cell protein, it is necessary to improve its accessibility to cellulase enzymes, destroying the lignin barrier surrounding cellulose fibers and to reduce the crystalline cellulose into amorphous form (2). In this research project is studied the effectiveness of different chemical and physical pretreatments and the enzymatic hydrolysis of cellulose. The saccharification of cellulose follows this schematic reaction pattern:

$$\text{Cellulose} \xrightarrow{C_1 + C_x} \text{Oligosaccharides} \xrightarrow{\beta\text{-Glucosidase}} \text{Glucose}$$

where C_x and C_1 are Endo and Exoglucanase respectively (3). Being the activity of C_1 and C_x strongly inhibited by cellobiose and that of -glucosidase by glucose (4), it is important to continously remove products of reactions from the reaction volume. In our study the enzymatic saccharification is performed in a membrane reactor. A membrane is selected which completely retains the enzymes and the macromolecular substrates, while the low molecular weight products are removed from the reaction volume. In such a way the product inhibition is controlled and a complete hydrolysis of the cellulosic substrate is possible.

This report, at the end of first year of the research program, is mainly concerning with the discussion of the following items:

i- optimal reaction conditions for the convertion of cellulose into oligosaccharides in a flat membrane reactor (of the stirred type)

ii- method for the determination of thermal stability of the different enzymes present in the cellulase complex

iii- effect of pretreatment on the enzymatic saccharification of olive wastes.

The present study fully covers points 1.1 and 1.2 of the proposal and partially points 2.2 and 4.

1.2 Results and discussion

All the experiments have been performed in a flat membrane reactor: volume 70 ml, membrane surface area 14.5 cm^2, membrane Amicon P.M. 10 type molecular weight cut-off 10.000. A detailed description of the apparatus used in this work is reported in (5). The hydrolysis of either commercial celluloses (Avicel, microcrystalline cellulose from Machery Nagel and CMC from Schuchardt) or native cellulose (olive-oil wastes, "sanse") has been carried out making use of Cellulase from Trichoderma viride (BDH, 0.02 EU per mg.).

The data have been worked out in terms of concentration of reducing groups (R.G.) and glucose (G) in the effluent cell stream. In Fig.1 are reported characteristic curves of $C_{R.G.}$ and C_G as function of process time. These curves allow the evaluation of most of the kinetic parameters. The maximum of the curve is characterized by the mathematical condition : dC/dt = 0, hence at the time at which the maximum occurs, the accumulation terms in the glucose and cellobiose material balances is zero, and the istantaneous reaction rates are given by the product of flow rate times effluent concentration. The measured concentration of reducing groups is the sum of glucose and oligosaccharides concentrations, hence it follows that:

$$r_1 = QC^{\wedge}_{R.G.} \qquad \text{and} \qquad r_2 = QC_G$$

where Q is the flow rate and $C^{\wedge}_{R.G.}$ is the effective concentration of oligosaccharides. In the meanwhile of performing via liquid chromatography a complete analysis of the effluent solutions, some preliminary very interesting results have been obtained with the following simple approach. Let suppose that oligosaccharides mixture is essentially cellobiose, in principle it is a mixture of cellodextrins (2-7 glucose monomers), if the solution is forced to react with excess β-glucosidase, the cellobiose would generate a double concentration of glucose. Indeed, under the experimental conditions adopted in the present study, it was found that, during Avicel and CMC hydrolysis, essentially cellobiose and glucose are produced, and only a very limited amount of other oligosaccharides are present in the reaction mixture. In fact glucose concentration in any sample, after reaction with excess β-glucosidase, increases and reaches the same value of the reducing groups. The concentration jump is equal to the double of the difference between reducing groups and glucose concentrations prior to the kinetic assay. This result allows to evaluate the rate of cellobiose production in the following way:

$$r_{Cell.} = Q\left(C_{R.G.} - \frac{1}{2} C_G \right)$$

The experimental data have been then plotted in terms of $1/r_{Cell.}$ vs $1/N_{Av}$, where N_{Av} are the mg of Avicel in the reaction volume, and $1/r_G$ vs $1/C_{Cell.}$ Both reaction rate are well represented by Michaelis-Menten rate equations with the following values of K_m : for cellobiose production K_m = 3.7 mg/ml and for glucose production K_m = .183 mM.Kinetic tests performed in a range of reaction temperatures between 25°C and 50°C enable the evaluation of an activation energy equal to 11.777 cal/moles for the first reaction synergistically catalized by C_1 and C_x (4) and an activation energy for β glucosidase equal to 10.855 cal/moles. These values are in fairly good agreement with those reported in Literature (6). Having been determined the rate equations and the activation energies of the reacting system it is possible to choice the optimal operational conditions.(see Figs. 2,3,4.)

A different set of experimental runs has been performed to independently test the deactivation with temperature of the different enzymes. All the experiments were carried on with a concentration of Trichoderma viride in the reactor equal to 0.033 mg/ml, and in the presence of δ-gluconolactone in the reaction mixture. The reactor is in a first case filled with amorphous cellulose (CMC 0.1% w/V) and fed with Na-acetate buffer solution 50 mM, pH 4.8, in order to test thermal stability of C_x. In a second case Avicel (33 mg/ml) is charged into the cell and the same buffer solution is continously fed. This experiment permits to evaluate thermal stability of the enzyme complex C_1, C_x. The presence of δ-gluconolactone inhibits the activity of β-glucosidase, thus avoiding the conversion of cellobiose into glucose. The thermal stability of β-glucosidase was determined feeding a 10 mM cellobiose solution in the same buffer to the reactor. Analysis of the experimental results indicates (see Fig.5) that the activity of C_x and C_1, determined by the rates of cellobiose production in the first two sets of experiments, is stable with process time, i.e. no thermal deactivation occurs, up to a temperature equal to 45°C, which is the maximum value investigated in this study. On the other hand β-glucosidase shows a thermal deactivation in the same range of temperatures. The activity of this enzyme is well represented by the following equation : $N = N_o \exp(-k_d t)$, where N and N_o are the istantaneous and initial (t=0) amount of active enzyme and t is the process time. At 45°C, the k_d value is $1.04\ 10^{-2}$ (1/hr), which implies that after 40 hours of working, the β-glucosidase has lost the 34% of its initial activity; hence it is necessary to immobilize β-glucosidase in order to drastically reduce its rate of deactivation. These results are confirmed by data of overall reaction rate, glucose production, obtained at different temperature (25-55°C) in long time experimental runs. In this case a saturating amount of Avicel (42.86 mg/ml) was charged into the reactor and the same buffer was continously fed, and glucose production was recorded as function of process time.

Finally experiments have been started to systematically investigate the effect of different pretreatments on the effectiveness of saccharification process. The results can be summerized as follows. A pretreatment of Avicel with Cadoxen (5% w/V Cadmium oxide in 28% acqueous Ethylenediamine) for 24 hr (8) and with H_2SO_4 solution (60 w%) at room temperature (1 hr, in order to avoid acid hydrolysis) determines an increase in the percentage of the amorphous form of cellulose and hence increases its accessibility to C_x enzyme. Cadoxen pretreatment is more active, in fact the concentration of reducing groups is higher as compared to that relative to H_2SO_4 pretreatment. Moreover in the case of Cadoxen flushing, the conversion of oligosaccharides into glucose is limited; this result indicates an insufficient amount of β-glucosidase in the complex and suggests the opportunity of a separate step of reaction performed in a second reactor using immobilized β-glucosidase. Acid and basic pretreatments of cellulose present in olive-oil waste were also carried on. The choice of olive-oil waste was suggested by their relatively large amount which can be disposed of in Italy. For further details, see (9). The NaOH and H_2SO_4 concentration in all the experiments was 1%w/V. The basic pretreatment destroys the lignin wall which surrounds the cellulose fibers, the acid pretreatment determines the conversion of hemicellulose to xylose. The results indicate that in order to get high conversion of the native cellulose it is necessary to make a basic pretreatment of cellulosic waste "sanse" for at least 3 hr. The lenght of acid pretreatment is almost ineffective on the degree of conversion. Few experiments have been at present time performed with cellulosic waste pretreated in two different ways : a) saturated steam at 150°C, b) Cadoxen at different times (4hr and 20hr). The results are encouraging, but further experiments are required in order to reach significant results.

Bibliography

1.1 N.M. Bikales, L. Segal "Cellulose and Cellulose derivatives", John Wiley and sons Eds., (1971).

1.2 M. Chang J. Polymer Sci. 36, (1971), 343.

1.3 E.T. Reese et al. J. Bacteriol., 59, (1950), 485.

1.4 D. Sternberg et al. Can. J. Microbiol. 23, (1977), 139.

1.5 G. Greco jr. et al Biotechnol. Bioeng. 21, (1979), 1421.

1.6 D.W. Sundstrom et al. Biotechnol. Bioeng. 23, (1981), 473.

1.7 F. Alfani et al. Enzyme Microb. Technol. submitted for publication.

1.8 M.R. Ladisch et al. Science 201, (1978), 743.

1.9 F. Alfani et al. 1st Int. Conf. on "Energy from Biomass", Brighton 4-7 November 1980.

Acknowledgements

The authors thank Dr. Donatella Albanesi and Dr. Amalia Vetromile for their assistance in the development of this research project.

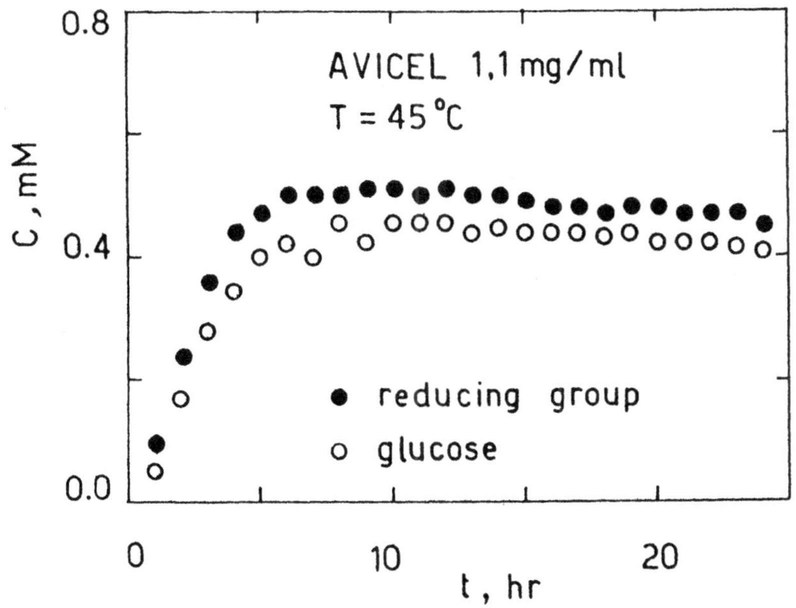

Fig. 1 - R.G. and G. concentration curves as function of process time

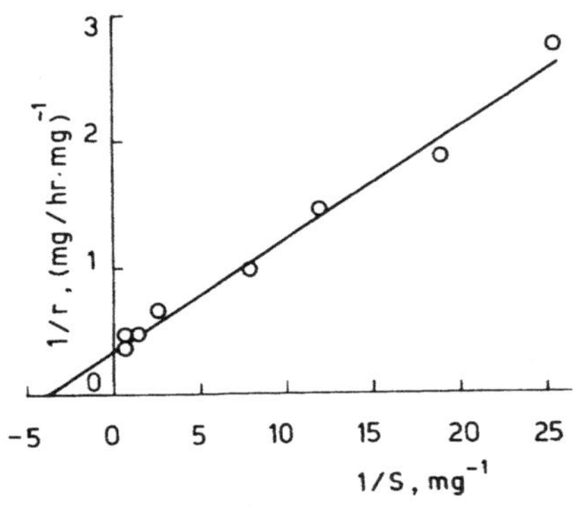

Fig. 2 - Lineweaver-Burk plot for $C_1 + C_x$ components at T = 45°C

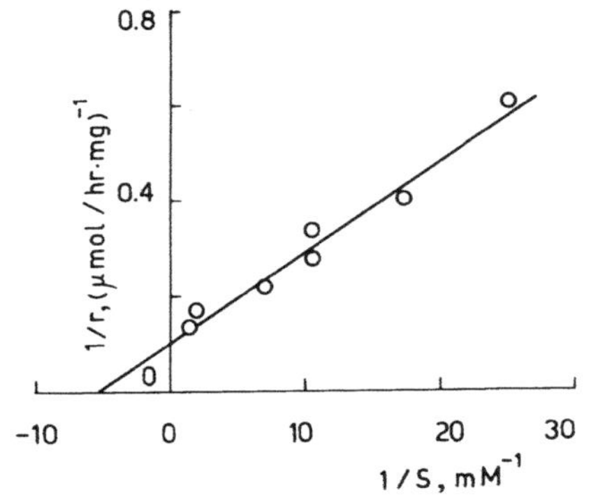

Fig. 3 - Lineweaver-Burk plot for β-glucosidase component at T = 45 °C

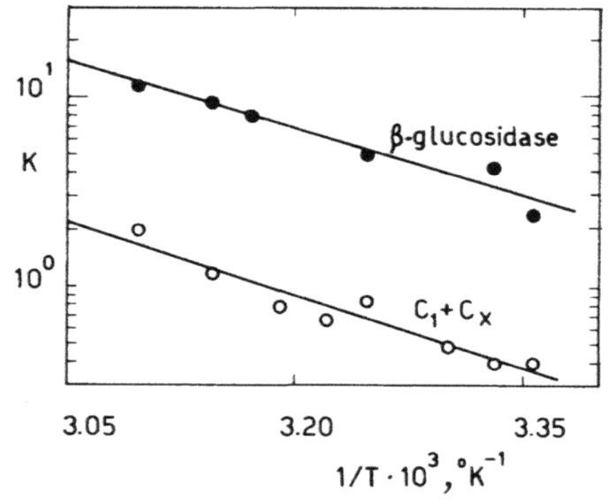

Fig. 4 - Arrhenius plot for the cellulase complex

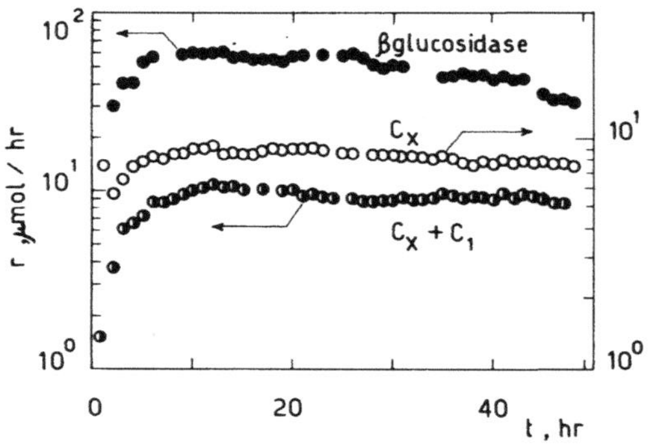

Fig. 5 - Thermal stability of the different components

SECTION 2

FINAL REACTION STAGE

The research work of Section 2 has been developed in close accordance with the program outlined in the previous coordination meeting held in Amsterdam on September 18th to 19th 1980.

Indeed, specific care has been devoted to the very final reaction step - i.e. to cellobiose hydrolysis by β-glucosidase. Two basic points have been investigated :

i) identification of optimal operating conditions in terms of temperature, pH and so forth in the membrane reactor;

ii) study of the stabilizing effects of partially hydrolized cellulose.

A mathematical model has been set up in order to predict the improvements in U.F. membrane reactor performance that stem from macromolecular substrates accumulation upstream from the membrane surface.

Preliminary experimental tests on a model system (dextrane hydrolysis by dextranase) have been undertaken.

HOMOGENEOUS PHASE β-GLUCOSIDASE KINETICS.

The tests have been performed with commercial β-glucosidase (BDH Ltd., Poole, U.K.). The results of a series of homogeneous phase kinetic tests have been reported in Fig. 2.1 in terms of reciprocal specific rate versus reciprocal substrate (cellobiose) concentration. The reaction temperature is 35 °C, all the runs have been performed at pH 4.80 in 100 mM Na-citrate buffer. By inspection of Fig. 2.1, it can be seen that β-glucosidase obeys Michaelis-Menten-type kinetics. The kinetic parameters are :

$$k_c = 5.15 \ 10^{-3} \text{ mmoles/mg min}$$

$$Km = 72 \text{ mM}$$

Glucose (product) concentration has been kept to a minimum because of enzyme inhibition phenomena observed at high conversion levels.

The activity versus pH curve for the homogeneous phase conditions is reported in Fig.2.2. The runs have been carried on at saturating cellobiose conditions (250 mM) in 100 mM Na-citrate buffer. The results have been worked out in terms of k_c/Km vs. pH for comparison purposes with the U. F. cell runs. The optimal pH value is 4.54.

The homogeneous phase thermal deactivation kinetics under saturating feed conditions have been also investigated. Typical experimental results have been reported in Fig.2.3 in terms of product (glucose) concentration versus reaction time. The shape of the curve suggests first order deactivation kinetics to hold with respect to active enzyme concentration. The data have been fitted with a non-linear least squares algorithm to produce the asymptote and the time-constant of the experimental curve. The latter parameter is equal to the reciprocal of the deactivation constant (k_d), the former being related to the kinetic constant of the enzymatic reaction.

The Arrhenius plot of the deactivation reaction is reported in Fig.2.4. The high activation energy ($3.45 \ 10^4$ cal/mol) suggests the process to actually consist in protein thermal denaturation (2.1).

As a by-product of the thermal denaturation runs, an Arrhenius plot for the main enzymatic reaction can be also determined. This is reported in Fig. 2.5. A value of $8.2 \ 10^3$ cal/mol has been determined for the activation energy which is in good agreement with previous results reported in the Literature (2.2).

DYNAMICALLY IMMOBILIZED ENZYME RUNS

Use has been made of an unstirred plane membrane cell (total U.F. membrane surface approximately 15 sq.cm, overall cell volume 30 ml) equipped with a DDS 600 cellulose acetate membrane (The Danish Sugar Co., Nakskov, Denmark), molecular weight cut-off of 20,000. Obviously, when operating with the whole cellulase complex, different membranes must be employed.

The experimental apparatus has been described elsewhere in detail (2.3). At the steady-state, the enzyme is confined (dynamically immobilized) within an extremely small volume immediately upstream from the membrane surface because of concentration polarization phenomena. Details of the enzyme concentration levels attained, that depend on the experimental conditions adopted, have been discussed elsewhere (2.4).

Typical experimental results, in terms of specific rate versus reaction time, have been reported in Fig. 2.6.

Preliminary experimental runs have been performed in order to identify the optimal pH value for the dynamically immobilized enzyme situation. In all the experimental runs, a non-saturating cellobiose concentration of 5 mM has been adopted. In this concentration region, the relationship between reaction rate and substrate concentration is linear and, hence, the ratio between the specific reaction rate and the corresponding substrate concentration is equal to k_c/Km. The results have been reported in Fig.2.2. It can be seen that an optimal pH of 5.15 holds. Hence, a shift in optimal pH occurs of more than half unit as compared to the homogeneous phase situation. This result is somewhat surprising since no manipulation has been performed on the enzyme, the only difference consisting in the considerably higher enzyme concentration levels attained in the dynamically immobilized enzyme situation. However, these yield considerable protein-protein interactions (the enzyme concentration is not far from the crytical value at which gel-precipitation occurs).

A second interesting feature of the experimental results consists in the enzyme activity levels being unaffected by dynamical immobilization.

Enzyme deactivation tests have been also carried out at pH 4.80 and at a constant cellobiose concentration in the feed of 5 mM. The same first order kinetics hold with respect to active enzyme concentration. This conclusion stems from log(specific rate) being linear with reaction time (see Fig.2.6).

In Fig. 2.4 the results have been reported in terms of deactivation constant k_d versus reciprocal absolute temperature. It can be seen that the activation energy of the enzyme deactivation is the same as the homogeneous phase and that a decrease in the deactivation constant k_d occurs at equal reaction temperature by a factor of approximately 3. The activation energy being unchanged as compared to the homogeneous phase situation, suggests that mass transfer resistances do not affect the phenomenon appreciably and hence that an actual enzyme stabilization has been achieved. This probably results from the same protein-protein interactions that can be credited for the observed optimal pH shift.

STABILIZED ENZYME RUNS

As already discussed in previous works (2.3-2.5), the injection of limited amounts of high molecular weight, water soluble, linear chain polymers into the completely polarized system gives rise to a high polymer concentration region in the same cell portion where the enzyme is dynamically immobilized. This results in considerable stabilization effects quite independently of the nature of both the enzyme and the stabilizing macromolecule because of drastic reductions in enzyme mobility due to the polymeric network that partially inhibit the unfolding of the enzymatic macromolecule.

Since high molecular weight, soluble cellulose fractions are obviously bound to be present in the hyrolysis reactor, the stimulating possibility exists that these could act as a stabilizer on the β-glucosidase performance. Experimental tests have been therefore performed by injecting into the system 15 mg of Carboxy-methyl-cellulose, in order to simulate, at least to some extent the actual system. Typical results appear in Fig. 2.6. It can be seen that a considerable stabilization has been indeed obtained at the expenses of virtually no enzyme activity decay. Indeed, the ratio between the homogeneous phase deactivation constant k_d and the one for the stabilized enzyme situation is equal to 22 at the same reaction temperature. Obviously this corresponds to an increase by the same factor in enzyme half life.

Further experimental tests are going on in order to identify the activation energy of the deactivation process for the stabilized enzyme situation.

Preliminary results seem to indicate that the optimal pH curve for the stabilized enzyme coincides with the dynamically immobilized enzyme one.

HIGH MOLECULAR WEIGHT SUBSTRATES

Both experimental and theoretical work has been started in the field of high molecular weight substrates conversion in polarized U.F. membrane reactors.

It is quite obvious that local increases in substrate concentration can be achieved at the active membrane surface if its molecular weight cut-off is such as to reject the macromolecular feedstock.

In a polarized reactor such as the one studied, these increases in concentration occur in the reactor region where the enzyme is confined. Hence, increases in specific reaction rate should occur as compared to a non-rejected substrate situation.

A mathematical model has been set up of the system in terms of a substrate mass balance equation that takes into account diffusion, convection and reaction according to a Michaelis-Menten rate equation. The actual enzyme concentration profile has been also considered.

The resulting non-linear differential equation has been solved numerically by means of a five points Runge-Kutta algorithm.

Typical theoretical predictions in terms of effectiveness Eta (actual reaction rate / reaction rate at a substrate concentration level equal to the feed one) versus a dimensionless group that accounts for the kinetic constant k_c, the enzyme amount $N°$, the permeate flow rate Q, the feed substrate concentration $C°$ have been reported in Fig.2.7 for rejection coefficient values of R=0.99 and R=0 respectively. It can be seen that, for diluted feedstocks, increases by a factor of approximately 20 in reaction rate can be achieved.

No experimental results are presented since some difficulties have arosen in the experimental work on a model system (dextrane hydrolysis by dextranase) mainly because of analytical problems.

BIBLIOGRAPHY

2.1 E.J. Carey, K.J. Leidler J.Am.Chem.Soc. 73, (1951), 1455.

2.2 D.W. Sundstrom, H.E. Klei, R.W. Coughlin, G.J. Biederman, C.A. Brouwer Biotechnol. Bioeng. 23, (1981), 473.

2.3 G. Greco jr., L. Gianfreda Biotechnol. Bioeng. 23, (1981), 801.

2.4 L. Gianfreda, G. Greco jr. 1st Int. Conf. on "Energy from Biomass", Brighton 4-7 November 1980.

2.5 L. Gianfreda, G. Greco jr. Biotechnol. Lett. 3, (1981), 33.

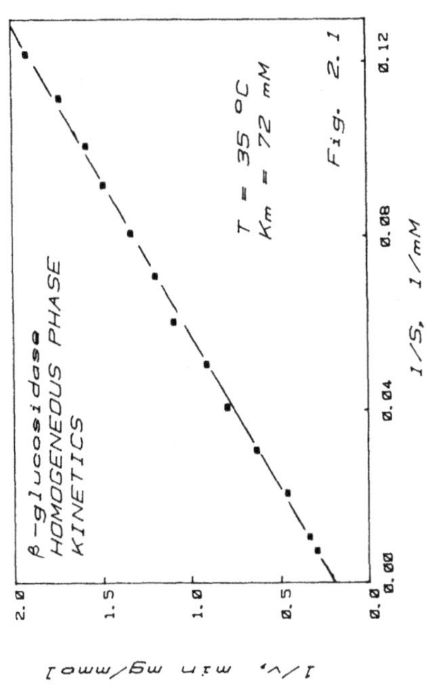
β-glucosidase
HOMOGENEOUS PHASE
KINETICS
T = 35 °C
Km = 72 mM
Fig. 2.1
1/v, min mg/mmol
1/S, 1/mM
2.0
1.5
1.0
0.5
0.0
0.00
0.04
0.08
0.12

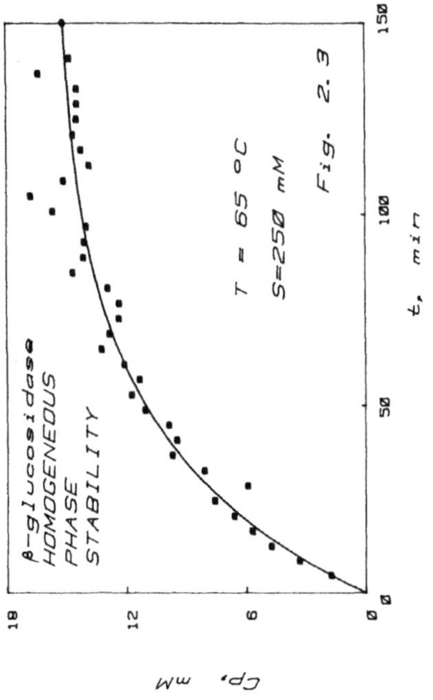
β-glucosidase
HOMOGENEOUS
PHASE
STABILITY
T = 65 °C
S=250 mM
Fig. 2.3
Cp, mM
t, min
18
12
6
0
0
50
100
150

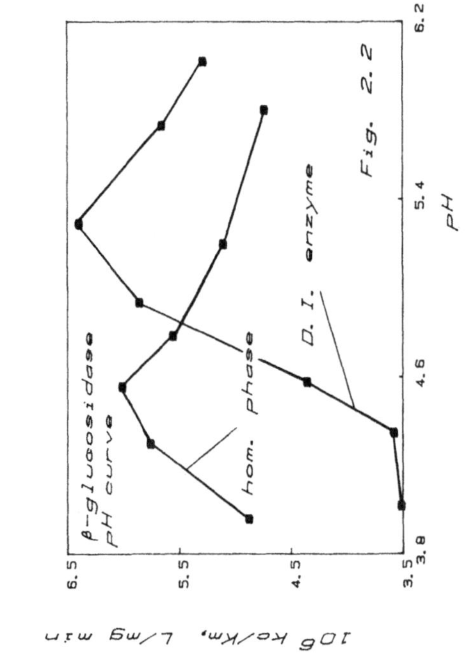
β-glucosidase
pH curve
hom. phase
D. I. enzyme
Fig. 2.2
10⁶ ko/Km, L/mg min
pH
6.5
5.5
4.5
3.5
3.8
4.6
5.4
6.2

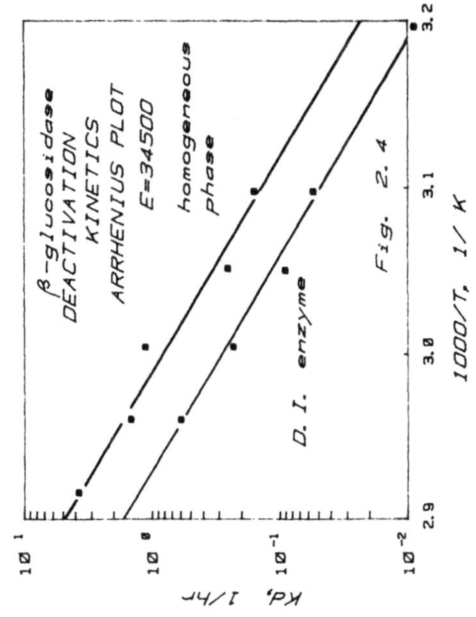
β-glucosidase
DEACTIVATION
KINETICS
ARRHENIUS PLOT
E=34500
homogeneous
phase
D. I. enzyme
Fig. 2. 4
Kd, 1/hr
1000/T, 1/ K
10 1
10 0
10 -1
10 -2
2.9
3.0
3.1
3.2

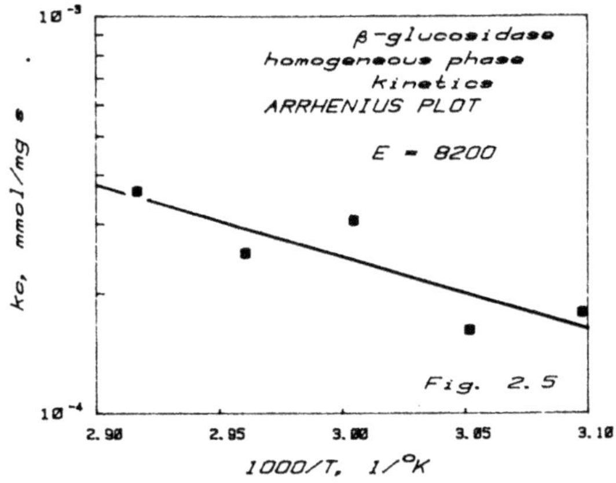
β-glucosidase
homogeneous phase
kinetics
ARRHENIUS PLOT
E = 8200
Fig. 2.5
kc, mmol/mg s
1000/T, 1/°K
10⁻³
10⁻⁴
2.90
2.95
3.00
3.05
3.10

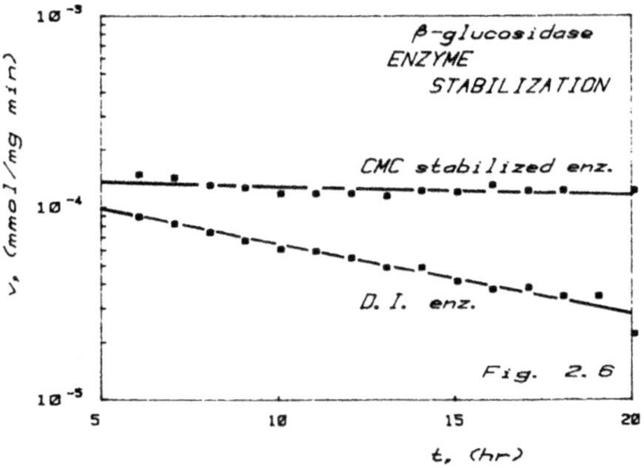
β-glucosidase
ENZYME
STABILIZATION
CMC stabilized enz.
D.I. enz.
Fig. 2.6
v, (mmol/mg min)
t, (hr)
10⁻³
10⁻⁴
10⁻⁵
5
10
15
20

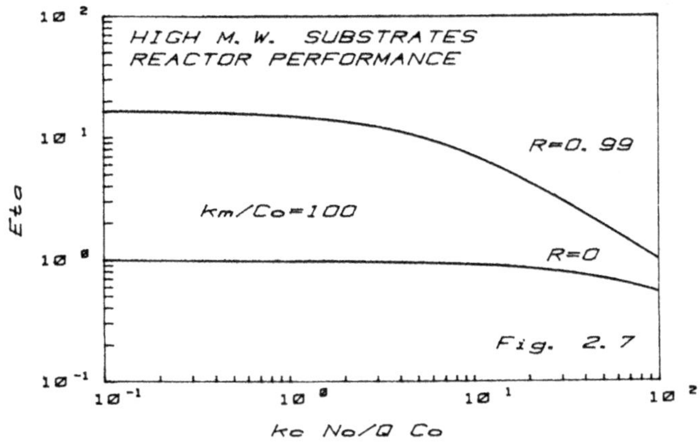
HIGH M.W. SUBSTRATES
REACTOR PERFORMANCE
R=0.99
km/Co=100
R=0
Fig. 2.7
Eta
kc No/Q Co
10²
10¹
10⁰
10⁻¹

LIQUEFACTION AND SACCHARIFICATION OF AGRICULTURAL BIOMASS

Authors : G. BELDMAN, A.G.J. VORAGEN, F.M. ROMBOUTS and W. PILNIK

Contract Number : ESE-R-038-NL

Duration : 36 months 1 July 1980 - 1 July 1983

Total Budget : 180.000 ERE CEC Contribution : 50%

Head of Project : Prof. Dr. W. Pilnik, Agricultural University, Department of Food Science

Contractor : Agricultural University

Address : Salverdaplein 10
P.O. Box 9101
6700 HB Wageningen
The Netherlands

Summary

In the bioconversion of solid agricultural biomass to energy carriers one can distinguish various stages: liquefaction, saccharification and fermentation to the desired fuel. Liquefaction and saccharification are indispensable if immobilized micro-organisms or enzymes are used for the fermentation step. Sugar-beet pulp which is a waste product of the beet sugar industry was chosen as a model for agricultural biomass. The pulp was treated with a combination of pectolytic and cellulolytic enzymes. This resulted in almost complete liquefaction and saccharification. The polygalacturonase from the pectic enzyme plays a major role during the liquefaction process; its optimal action is at pH 3,5. The solubilized neutral sugars were in monomeric form. The sugar solution consisted of glucose, fructose, arabinose and galacturonic acid. Cellulose was hydrolysed extensively: 91% was converted to glucose. The hemi-cellulose fraction, which consists mainly of araban was hydrolysed for 93%. From the pectic substances 91% was brought into solution. Residual saccharose of the pulp was converted to glucose and fructose. The obtained sugar solution is a versatile product and can be used for subsequent conversion to several fuels like ethanol, hydrogen or methane.

I.1 Introduction

Solar energy can be used as a renewable energy source through photosynthesis by green plants. About 0.1% of the solar energy reaching the earth, is fixed in about 10^{11} tons organic plant substance per year (1). Approximately 40-50% of this material is cellulose, an important part of the plant cell wall. Next to cellulose the plant cell wall consists of pectins, hemi-cellulose, lignin and structural proteins.

The conversion of cell wall material to energy carriers goes through several stages. Firstly the liquefaction of insoluble cell wall polysaccharides, secondly the saccharification of the solubilized polysaccharides to single sugars and thirdly the fermentation of the sugars to energy carriers like ethanol, methane or hydrogen. Liquefaction of solid material like plant cell walls is an important phase in the bioconversion and indispensable for fermentation in systems with immobilized enzymes or immobilized micro-organisms.

The bioconversion of the cell wall material requires a large number of enzymes. The accessibility of the substrates is a restrictive factor for enzyme action, in particular this is the case for cellulose. The degradation of cellulose in several fruit and vegetable tissues is enhanced by simultanious hydrolysis with pectic and cellulolytic enzymes (2,3).

This report describes the enzymatic liquefaction and saccharification of sugar-beet pulp, a waste product of the beet sugar industry. In the Netherlands 5 x 10^6 tons of pulp per year are produced. Processing to cattle feed requires an amount of energy (in the Netherlands 1.5 x 10^8 m^3 natural gas per year), which makes the product less profitable. Enzymatic hydrolysis to a sugar solution may be an alternative process and may open the way to use this material as a raw material for the production of energy.

I.2 Materials and Methods

Sugar-beet pulp was from the sugar plant of the Suiker Unie, Puttershoek, the Netherlands. Pulp samples were taken from the diffusion tower and stored at -20°C. Enzyme preparations were from Gist-Brocades, Delft, The Netherlands. The cellulolytic enzyme -Maxazyme CL 2000- is produced by the fungus *Trichoderma viride*. The pectolytic enzyme -C80- is produced by the fungus *Aspergillus niger*.

Liquefaction experiments: Liquefaction experiments were carried out in the thermostated rotating reaction vessel of a Brabender Viskograph (Brabender, OHG, Duisburg, G.F.R.) at 40°C and 100 r.p.m. The pulp was ground in a meatgrinder (Kenwood, Chef A700). To be in the resistance range which can be measured by the instrument, 70 g water had to be added to 100 g pulp. The particle size was further reduced with an Ultra Turrax (X-1020 Intern. Laborat. App. Dottingen, G.F.R.). After temperature equilibration 15 ml of a solution containing 0.2 g Maxazyme CL 2000 and 0.25 g C80 was added. For measurement of consistency changes, the strain exerted on a body, placed in the rotating reactor was recorded as a function of the reaction time and expressed as Brabender Units (B.U.).

Enzyme assays: Exo-β-1,4-glucanase or C_I activity was measured by the increase of reducing end groups released from crystalline cellulose (Whatman no.1 filter paper, pieces of 2 x 2 cm; 33,2 mg) in 0.1 M sodiumacetate buffer pH 4.0 and 40°C. The quantity of enzyme which produced the equivalent of 3.32 mg glucose (10% hydrolysis) after 2 hours was determined. Specific activity was expressed as units/g enzyme (one unit is 1 µmole reducing end groups released per minute). Endo-1,4-β-glucanase (C_x), cellobiase, polygalacturonase, pectin lyase, galactanase, arabanase and pectinesterase activitties were measured according to Voragen *et al.* (4).

Sugar analysis: Reducing end groups were measured according to Nelson-Somogyi (5). The determination of the total sugar content was done by the method of Dubois *et al.* (6). The galacturonide content was estimated with the method of Ahmed and Labavitch (7). Neutral sugars were analysed gaschromatographically as alditol acetates (4) and by HPLC using a LiChrosorb NH_2 column.

I.3 Results

Activities of the commercial enzyme preparations: All commercial enzyme preparations available to us are investigated on their liquefaction capability. All combinations of pectolytic and cellulolytic enzymes which contain the complete cellulase complex (C_I, C_x and cellobiase) and a broad pectic enzyme spectrum give almost complete liquefaction. In our further work we used Maxazyme CL 2000 combined with C80. Extra addition of hemicellulase and cellobiase to this enzyme combination does not improve the liquefaction process, so the Maxazyme and C80 preparations contain enough of these enzymes. Table I shows the activities of various enzymes present in the Maxazyme CL 2000 and C80 preparations. Pectolytic activity is not found in the Maxazyme CL 2000 preparation. Exoglucanase activity is only found in Maxazyme CL 2000, while endoglucanase and cellobiase activity appears in both enzyme preparations. Arabanase and β-1,4 galactanase activities were mainly found in C80.

Chemical analysis of the material: The chemical composition of sugar-beet pulp is presented in Table II. The dry weight content of the material is 9,75%. The unpressed pulp still contains a considerable amount of saccharose. The hemicellulose fraction consists mainly of arabinose.

Viscosimetric studies in the Brabender Viscograph: The decrease in viscosity of the sugar-beet pulp is shown in Fig. 1. During liquefaction the pH, the released reducing sugars and the residue of the water insoluble solids are measured and shown in the same figure. The pH shows a decrease from 5,4 to 3,8. At the lower pH (after 2 to 4 hours incubation) the process of liquefaction is accelerated. Fig. 2 confirms this pH effect. If the initial pH of the pulp is brought to 3.8 with HCl, the initial decrease of viscosity is much faster. Experiments in which the cellulolytic and pectolytic enzymes are added to the pulp separately, at this low initial pH, show that pectic enzymes are responsible for this effect. The optimum for the initial pH is determined by measuring the decrease of viscosity during the first 15 minutes of liquefaction for several pH values. This optimum is at pH 3.5 while the pH optimum for polygalacturonase in the pectic enzyme preparation is pH 3.8 (Fig.3). The results suggest that polygalacturonase is responsible for the fast initial decrease in viscosity. Although the pH optimum of the cellulase complex is at pH 4,8 the enzyme did not lose activity at this lower pH.

Analysis of the reaction products: The degree of polymerization of the sugars, released by enzymatic hydrolysis during 22 h was examined by gel filtration on Biogel P2. Fig. 5 shows that almost all of the neutral sugars are present in monomeric form. The uronic acids appeared in two different peaks. The void contains high molecular weight pectin fragments, while fractions 90-99 contain oligomeric fragments.
Table III shows the composition of the reaction product. Neutral sugars were analysed by HPLC. The composition of the residue remaining after hydrolysis is given in Table IV. A small part of the polysaccharides is not attacked by the enzymes and found in the residue.The percentage hydrolysis of the different (poly)saccharides is given in Table V. Saccharose was converted to glucose and fructose by the action of invertase, belonging to the pectic enzyme. The cellulose which is a very resistant part of the material is extensively hydrolysed.

I.4 Discussion

The bioconversion of agricultural biomass for the production of alternative fuels is a promising process. Production of fuel alcohol from starch seems to be commercially feasible but some disadvantages must not be overlooked. The use of nutritional starch for the production of fuels can cause some ethic and environmental objections like raising food prices and overuse of pesticides and fertilizers. The use of agricultural waste in its place may overcome these problems (8). Solid wastes can be digested anaerobically to the fuel methane, but a more versatile product can be obtained if the waste polysaccharides are hydrolysed to fermentable sugars. In this paper we have shown that sugar beet pulp can be converted to a monomeric sugar solution, with pectolytic and cellulolytic enzymes. There is a synergistic effect in the action of the two enzyme preparations. The polygalacturonase plays a major role in the liquefaction, its optimal activity is at pH 3.8. At this pH the cellulolytic enzymes are still very active.

The process of liquefaction and saccharification to a sugar solution is extremely useful if immobilized enzymes or micro-organisms are applied for subsequent conversion.

The use of sugar beet pulp as a waste with a low lignin content, gives high yields in cellulose hydrolysis (90% yield). If waste material with a high lignin content is used, one should take into account a much lower cellulose conversion (40-50% yield) and higher enzyme and pretreatment costs (8). Other wastes, which in composition can be compared to sugar beet pulp and can be used in this process are listed in Table VI (9). For the production of fermentable sugars from sugarcane simultaneous enzymic hydrolysis of cellulose and sucrose is possible (10). In order to produce ethanol from whole sugar beets as an energy crop, the process described here may be an improvement. Sucrose and cell wall polysaccharides can both be hydrolysed to fermentable sugars. Other energy crops may be used in the same way.

In order to make an economically feasible process our future research will be focussed on developing a continuous process.

Efforts will be directed to optimalize enzyme concentrations and to study possibilities to recover the enzymes by binding to new substrate or by applying ultrafiltration techniques. Also, experiments with isolated enzymes may give more information about the nature (stability, substrate affinity, pH optimum) and function of the enzymes needed for liquefaction. Finally, we will try to improve the process by certain pretreatments of the material.

II. Conclusions

It is possible to liquefy and saccharify solid agricultural waste to a fermentable sugar solution. As a model sugar-beet pulp is hydrolysed with a combination of pectic and cellulolytic enzymes. We conclude: 1)Polygalacturonase plays an important role during liquefaction; 2) The optimum for the process is at pH 3.5; 3) The solution obtained contains mainly glucose, fructose, arabinose and galacturonic acid; 4) The neutral sugars are in monomeric form; 5) The polysaccharides are hydrolysed for a great extent: cellulose for 91%, araban for 93% and pectin for 91%.

References

1. Ruy, D.D.Y. and Mandels, M., Enzyme Microb.Technol., 2, 91-102(1980).
2. Pilnik, W., Voragen, A.G.J. and de Vos, L., Flüssiges Obst, 42, 448-451 (1975).
3. Voragen,A.G.J., Krist,R., Heutink,R. and Pilnik,W., Proc.II.Intern. Congr. on Engineering and Food, Helsinki, Finland (1979).
4. Voragen,A.G.J., Heutink,R. and Pilnik,W., J.Appl.Biochem. (in press).
5. Spiro,R.G., Methods Enzymol., 8, 3-25 (1966).
6. Dubois,M., Gilles,K.A., Hamilton,J.K., Rebers,P.A. and Smith,F., Analytical Chem., 28, 350-356 (1956).
7. Ahmed, A.E.R. and Labavitch, J.M., J.Fd Biochem., 1, 361-365 (1977).
8. Mandels, M., ASM News, 47, 174-178 (1981).
9. N.R.L.O. Report Agrification, The Hague (1981).
10. Jones, C.K. and Yang,R.Y.K., Chem.Eng.Commun., 6, 283-291 (1980).

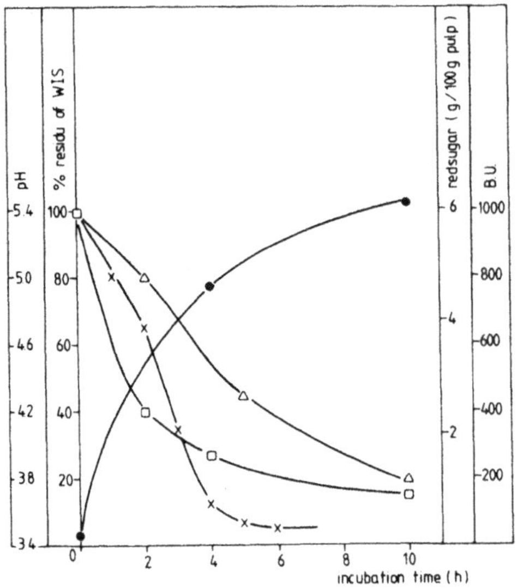

Figure 1.

Decrease of viscosity (x), pH (□), residue of water insoluble solids (△) and increase of released reducing sugars (●), by the combined action of pectinase and cellulase preparations.

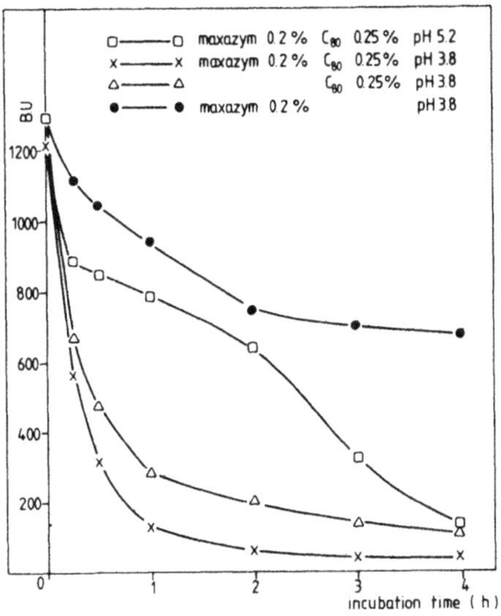

Figure 2.

Decrease of viscosity of sugar-beet pulp, treated with pectinase, cellulase and a combination of these enzymes at different pH's.

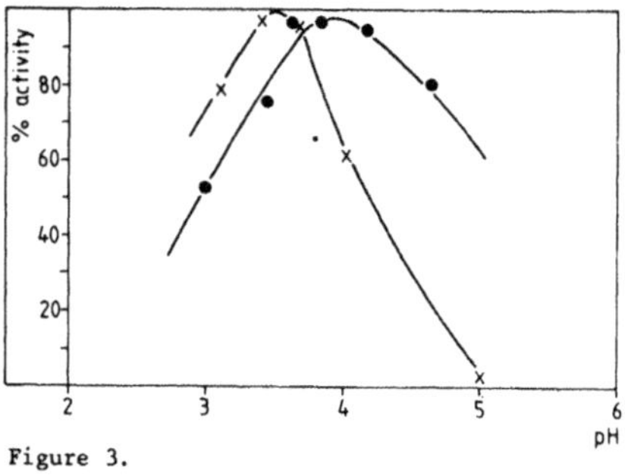

Figure 3.

Optimum pH for liquefaction, measured in Brabender Viscograph (X) and optimum pH for polygalacturonase (●).

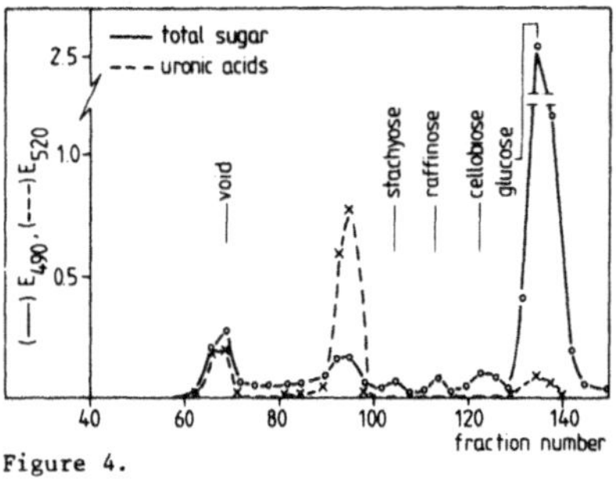

Figure 4.

Gel filtration on Biogel P2 column of fragments released from sugar-beet pulp.

Table I

Specific activity of commercial enzyme preparation (units/g)

Enzyme	Maxazyme CL 2000	C80
exo-1,4 glucanase	39.3	-
endo-1,4 glucanase	750.0	344.0
cellobiase	32.4	120.0
pectin lyase	-	50.4
pectinesterase	-	767.0
polygalacturonase	-	421.0
arabanase	0.5	44.4
galactanase (β-1,4)	0.4	37.5

Table II

Composition of sugar beet pulp

	dry weight %	
Sugar from polymers		48.0
Arabinose	16.0	
Xylose	0.9	
Galactose	3.2	
Glucose	14.9	
Galacturonic acid	13.0	
Saccharose		30.8
Protein		5.1
Ash		9.0
Lignin		2.0
Total		94.9

Table III

Composition of enzyme digest

	g/100 g pulp (wet weight)
Glucose	3.66
Fructose	1.56
Arabinose	1.63
Galacturonic acid	1.44
Total	8.29 = 85% of dry weight

Table IV

Composition of pulp residu after enzymatic hydrolysis

	dry weight %
Arabinose	6.0
Xylose	2.6
Galactose	1.5
Glucose	8.6
Galacturonic acid	6.5
Protein	17.5
Ash	18.3

Table V

Hydrolysis limits of various (poly)saccharide fractions

	% hydrolysis
Cellulose	91
Araban	93
Pectin	91
Saccharose	100

Table VI

Some agricultural wastes produced in the Netherlands (tons/year)

	wet weight	dry weight
auction surplus:		
vegetables	51,000	3,150
fruits	21,000	2,800
canning waste	62,000	6,500
onions:		
sorting waste	60,000	48,000
eventually		
horticultural plant waste		
tomato, cucumber plants	35,000	6,000
flour plants	81,000	14,000
flour bulb waste	96,000	8,800

SUBJECT V. BIOMASS CONVERSION (THERMOCHEMICAL ROUTES)

THE DEVELOPMENT OF FURNACE/HEAT EXCHANGER SYSTEMS IN WHICH CHOPPED CEREAL STRAW IS THE FUEL

Contract Number	:	ESER 026 UK(N)
Duration	:	1 July 1980 - 30 June 1983
Total Budget	:	£43,000 CEC Contribution : £21,500
Head of Project	:	B. Wilton, Faculty of Agricultural Science, Sutton Bonington, Loughborough, Leics., England
Contractor	:	University of Nottingham
Address	:	University Park, Nottingham

Summary

Much of the current straw burning equipment for farmhouse heating is unsatisfactory in several respects - for example manual bale handling, batch fuelling and combustion in a water-cooled chamber. Attempts to overcome these problems, most of which stem from the fact that the straw is harvested in bales, tend to involve considerable expense.

In many ways a more logical approach is to harvest, store, handle and burn straw in the chopped form, but some of the proposals for such a system would be expensive to install and would have the disadvantage of storing the straw adjacent to the dwelling and very close to the boiler. An approach which should be cheaper, cleaner and safer is to use a totally enclosed trailer to act as a mobile hopper/stoker and in order to improve the efficiency of combustion, to use a furnace/heat exchanger unit rather than a boiler.

1.1 Introduction

It is only in very recent years that straw has begun to be regarded as a potential fuel in the United Kingdom and initially the equipment used to burn straw was all imported, mainly from Scandinavian countries. There are now a few UK manufacturers of straw-fired boilers however, and their early equipment was very similar to the imported units. Recently some original designs are beginning to appear.

The early boilers were mainly designed to take either one or two small bales of straw : they were cylindrical, were refuelled by opening a large end door when the fire had died down, and the combustion chamber was surrounded by a water jacket. The consequences of batch firing, combustion in a water-cooled chamber and regulation by controlling the amount of air available for combustion were relatively low efficiencies, uneven output and considerable amounts of smoke. Despite these drawbacks ever-larger versions of these units have been manufactured and some that are used to heat glasshouses will hold several large cylindrical bales per charge.

Improvements in efficiency have been achieved by moves towards continuous firing. On small units this has meant the installation of holding conveyors or magazines for small bales, together with a chopper to reduce the straw to a form in which it can be readily conveyed into the combustion chamber. In some cases the latter is now a firebrick-lined 'retort', the former 'boiler' now being known as a 'heat exchanger'. Some improvement in efficiency has been achieved, however the complexity of the system is greater and the capital cost is considerably higher. There are chopper/feeder units now available that are capable of receiving large cylindrical bales, and for large units it is possible to stoke on demand with single whole small bales from a conveyor (although in the latter system broken and deformed bales cause problems which have yet to be overcome).

As far as is known all the present UK installations operate on baled straw, with or without a shredder, even though it can be argued that it is much more logical to operate on chopped straw if the storage and handling problems can be overcome. Although there was originally considerable enthusiasm for straw burning, this now seems to have declined. The newer, more complicated units are rather expensive and a number of the earlier, less efficient ones are not now being used by their owners because of the inconvenience of stoking and problems with smoke and smuts.

At Nottingham a large furnace/heat exchanger unit has been developed, designed to operate in conjunction with a grain drier on chopped straw, produced as a by-product of a whole crop cereal harvesting system. Surplus straw from this system would be available for domestic heating (Figure 1); alternatively it would be possible to gather straw from combine harvested crops and store it in the chopped form for later use, as outlined by Persson and Have (1).

1.2 Experimental Work

Although the work done in Denmark provides one solution to the problem of handling straw it could involve considerable capital expenditure - in storage buildings for chopped straw, in a blower to fill this store and in sophisticated equipment for collecting the straw from the store and transferring it on demand to an adjacent heating unit. Apart from the expense there are three aspects of this system which it is believed would restrict its uptake in the UK. These are:

(i) the inherent danger of having the chopped straw and the boiler so near to one another;

(ii) the need to locate the straw store near the farm house to minimise the length of the hot water feeder pipe;

(iii) the dust which is inevitably associated with straw, particularly when it is handled pneumatically.

To overcome these problems and make chopped straw more attractive as a fuel in the UK (where farm dwellings are usually not immediately adjacent to farm buildings) an alternative approach is being developed at Nottingham. This is based on one or more stores, which may or may not be in buildings, and a totally enclosed trailer which will be used to transfer straw from a store to a furnace/heat exchanger unit. The latter can be located in a garage-sized building attached to the building to be heated.

The trailer (Figure 2) will be moved and filled by a tractor fitted with a loader; it will then be parked and act as a hopper/dispenser for the furnace and will hold one day's supply of straw. The trailer has been designed with a hinged lid to eliminate the dust emission problem, and the lid carries a suspended agitator to encourage the straw to fall into the discharge auger. An auger has been chosen so that the spillage and carry-over problems of the more common slatted and belt floor self-unloading trailers can be avoided. The parked trailer will be plugged in to the furnace's electrical control system and will carry two intermittently operating electric motors.

The furnace (Figure 3) will be a scaled-down version of the sloping grate unit referred to earlier. The grate is composed of trimmed fire-bricks mounted on an adjustable frame, this design replacing the water cooled grate on the larger unit to avoid the need to dispose of, use or cool the heated water. The whole feeder/grate unit will be on skids to aid removal for inspection, modification and repair. A side door is provided to allow some other solid fuel to be placed in the unit to cover the trailer refilling time (which will typically be some 15 minutes).

A 2-pass heat exchanger will be carried on legs and will straddle the furnace. Movement of flue gases through to the chimney will be fan assisted, although at low loads induced draught should be sufficient to maintain combustion.

x x x

It is thought that the combination of a simple store (which may even be a field heap), an enclosed trailer/feeder hopper and a controllable feed firebrick furnace should lead to a clean and efficient system which will be completely mechanised.

Reference

(1) Persson, K. and Have, H. Handling and storage of chopped straw for heating pruposes. CEC project no. 324-78-ES-DK, (1980)

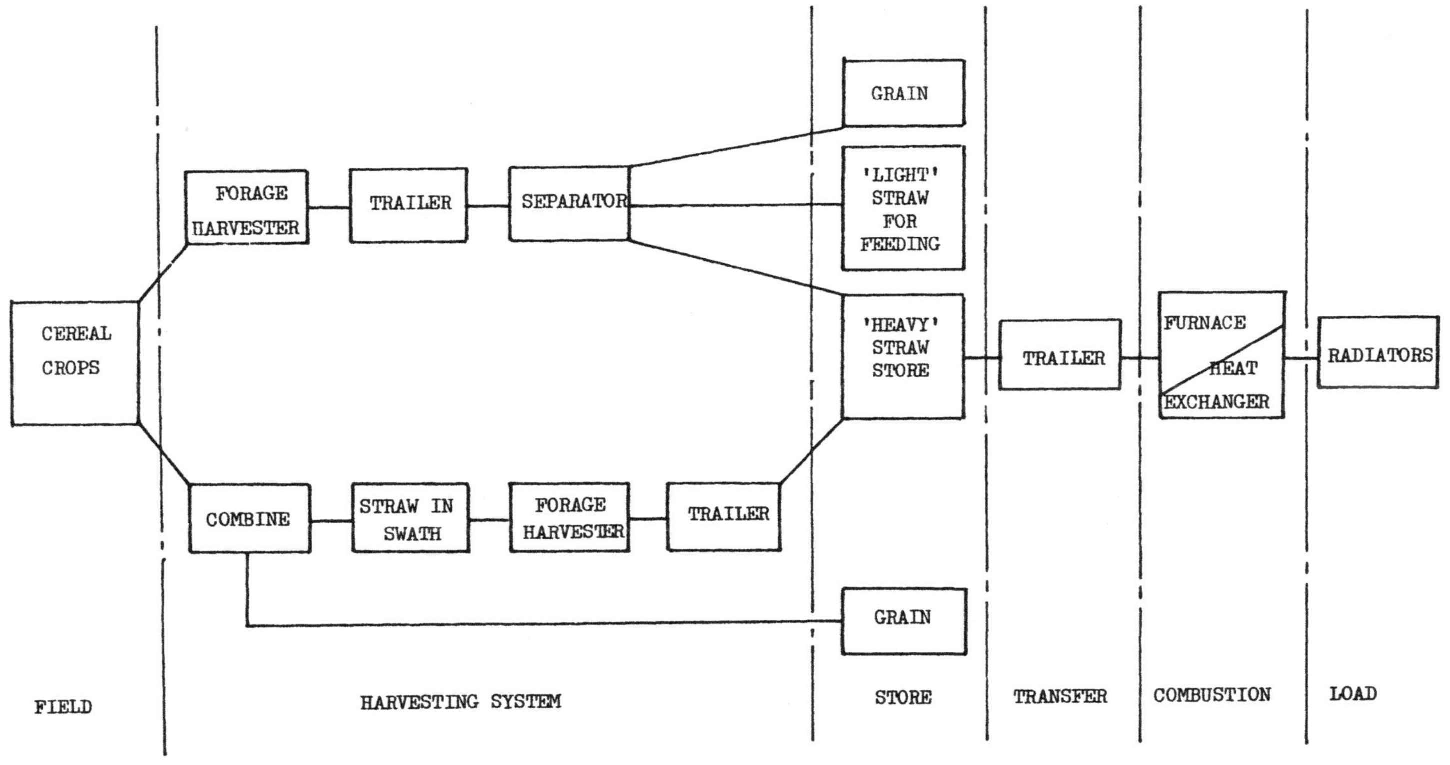

FIG. 1 TWO SYSTEMS OF PROVIDING CHOPPED STRAW FOR FUEL

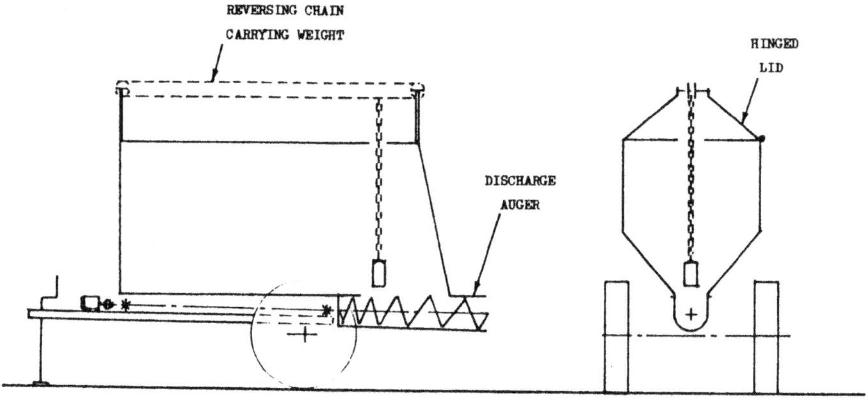

FIG. 2 THE ENCLOSED SELF-EMPTYING TRAILER FOR CHOPPED STRAW

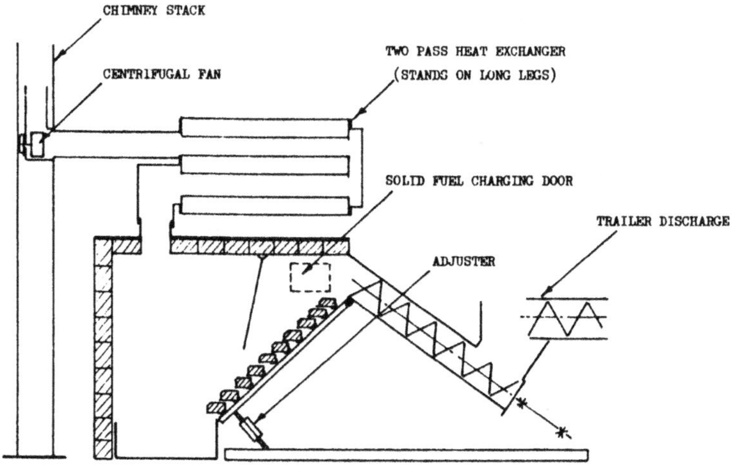

FIG. 3 THE FURNACE/HEAT EXCHANGER UNIT

THE USE OF GAS SCRUBBERS FOR HEAT EXTRACTION FROM STRAW FURNACES

Authors : H. Kofoed Nielsen and H. Have

Contract number: ESE-R-027-DK (G)

Duration : 36 months 1 July 1980 - 30 June 1983

Total budget : Dkr. 1.571.710 CEC Contribution 50 %

Head of Project: Prof. T. Tougaard Pedersen

Contractor : Jordbrugsteknisk Institut
Royal Veterinary and Agricultural University

Address : Jordbrugsteknisk Institut
Royal Veterinary and Agricultural University
Rolighedsvej 23
DK-1958 Copenhagen V

Summary:

The purpose of the project is to improve the efficiency of straw furnaces by cooling down the flue gas below dew point and thus making the upper calorific value of straw a realistic target for energy utilization. The heat exchanger to be used is a gas scrubber, which also is expected to reduce the emission considerably.

The paper present results from basic experiments with gas scrubbers used as heat exchangers. The aim of the work has been to investigate various packing elements and to determine constants being necessary for designing in actual cases.

The results of the experiments show that the gas scrubber is a very efficient heat exchanger. For example it has been calculated that an 18 kW heat loss through 300°C hot flue gas from a straw furnace can be recovered by a scrubber having a diameter of 230 mm and a height of 120 mm.

1. Introduction.

Straw furnaces of simple design often work with very low efficiency because of high flue gas temperatures. The aim of the project is to develope a gas scrubber which can be used as supplementary heat exchanger in connection with such straw furnaces, or as the sole heat exchanger in furnaces without the usual water jacket. By this method utilization of the upper calorific value becomes a potential possibility. Therefore very high thermal efficiencies can be expected.In addition a significant reduction of emission is expected because the flue gas is cleaned in the scrubber.

During the first year of the study attempts have been made to produce a basis for design of gas scrubbers for the purpose. Information from literature and from producers of packing material is not applicable because the operation conditions differ much from those used in cooling towers. Gas scrubbers which are to be used as heat exchangers usually must operate with much higher temperature differences and lower gas and water flows. Therefore it has been necessary to carry out some basic experiments to determine the lacking constants and to get experience with the ranges of water and gas flows for various packing materials.

In the following the first results from these experiments are presented.

2. Description of apparatus and measuring equipment.

The measurements are carried out on the experimental scrubber sketched in fig. 1. The packing material is placed in a steel cylinder, which has the lower end in a tank with water. Air is delivered to the scrubber by a fan through an 11 kW variable electric heater which is able to heat up the air to a maximum of 700^oC. Water is delivered to the packing material from the tank by a pump through a nozzle. Having passed the packing material it is collected in a small container where the temperature is measured. The water in the tank is cooled by a cold water spiral.

The diameter of the scrubber was determined on basis of the theory of cooling towers, (1) and (2), the available air flow and temperature range and the desired temperature raise of water. It was found that a diameter of 150 mm was suitable. The height, which was to be determined in the experiments, was made changeable by dividing the packing column into sections.

The filling elements tried so far are 3/4" ceramic intalox saddles and 5/8" metal pall rings. Testing has been carried out with air flows in the range 8-30 g/s and air temperatures in the range 80-450^oC.

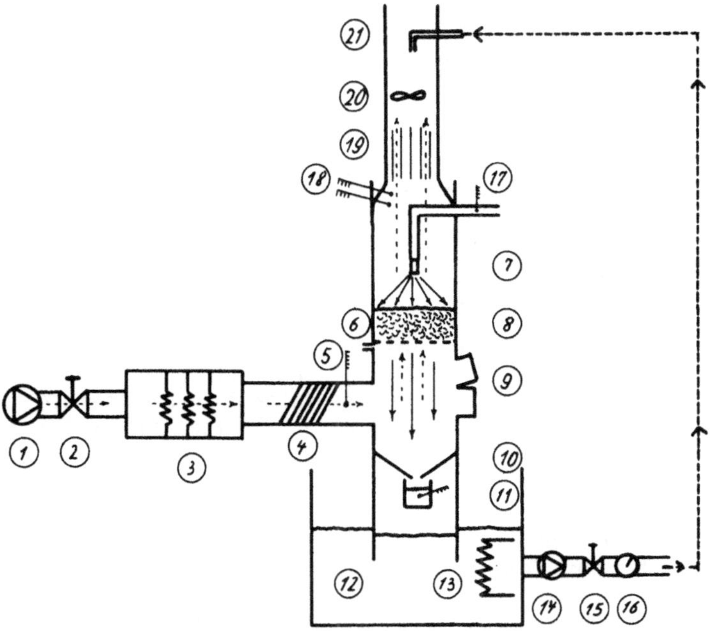

Fig. 1. Sketch of experimental scrubber.

1.	Variable fan.	12.	Water container
2.	Valve	13.	Heat exchanger for cooling
3.	11 kW variable electric heater	14.	Pump
4.	Augers for air mixing	15.	Valve
5.	Thermo couple	16.	Flow meter
6.	Pipe for measurement of static air pressure	17.	Thermometer
		18.	Psycrometer
7.	Nozzle	19,20	Anemometer
8.	Packing elements	21.	Flowmeter
9.	Windows	→	Water flow
10,11	Temperature measurement, washing water	--→	Gas flow

3. Results.

The first experiments were carried out with a 3 m high packing column. The measurements showed that the temperature of air flowing out of the scrubber was very close to the temperature of water flowing in and that the difference between the two temperatures was independent of the temperature of air going to the scrubber.

These results indicated that the packing column was higher than required for a satisfactory efficiency. The measurements were therefore repeated with a 500 mm high packing column. But the results obtained were similar to those found for a height of 3 m. The conclusion was therefore that also a 500 mm packing column was higher than required.

More series of measurements with packing columns were then carried out with heights of 250, 150 and 50 mm. The results from those showed that the height now was at the right magnitude, but even the 50 mm packing height gave a reasonable efficiency. However the distribution of water over the packing elements was very crucial for the performance of this height.

The results obtained have been used to determine the overall mass transfer coefficient from the theory of cooling towers (2). This constant make it possible to design scrubbers for actual purposes. The accurateness of the constant determined is however limited because of minor problems with the air flow measurements. It should therefore only be used as a first approximation.

Concerning the packing elements it was observed that the degree of wetting was very important for the efficiency, although the results seem to show, that it is possible to use less water than recommended for cooling towers. Of the two tested packing materials the 3/4" ceramic intalox saddles seem to be the best.

4. Analysis.

The results obtained so far show that the gas scrubber is a very efficent heat exchanger. This can be illustrated by the following example:

A typical manually stoked straw furnace is fired with 10 kg straw/h and operates with an air surplus of 4 times the theoretical (62 g/s). The flue gas from the furnace has a temperature of 290°C and an enthalpy of 396 kJ/kg.

The furnace is assumed fitted to a gas scrubber which cools down the flue gas to 30°C (enthalpy 99 kJ/kg). The water to the scrubber is supplied at 20°C and is heated up to 50°C. The power transmitted from the flue gas to the water is 18,3 KW.

The size of the scrubber can be calculated by application of the theory of cooling towers and use of the constant found in the experiments. This calculation shows that a scrubber having a diameter in the range of 150-230 mm and a height of 120 mm would work with an efficiency of approximately 88 % (96 % of lower calorific value). The preliminary conclusion is therefore that the capacity of a gas scrubber operating as a heat exchanger is in the magnitude of 4 - 9 MW/m^3.

5. Conclusion.

The paper describes some experiments with a gas scrubber used as a heat exchanger for transmission of heat from 250-300°C hot air to 20-50°C hot water. The main results are:

1. It is possible to scrub and cool the flue gas with gas scrubbers designed according to the common theory of cooling towers.
2. Gas scrubbers are very efficient heat exchangers. The capacity is in the magnitude of 4-9 MW/m^3.

6. References.

(1) Hansen, L.A. and Søltoft, P. (1980): Kemiske enhedsoperationer. - Akademisk Forlag.

(2) Gurney, J.D and Cotter, I.A. (1966): Cooling Towers. - Maclarn & Sons Ltd.

HEAT ENERGY FROM ANIMAL WASTE BY COMBINED DRYING, COMBUSTION AND HEAT RECOVERY

Authors : H. Have and M. Fritze

Contract Number: ESE-R-028-DK (G)

Duration : 36 months 1 July 1980 - 30 June 1983

Total Budget : Dkr. 1.886.220 CEC contribution 50 %

Head of Project: Prof. T. Tougaard Pedersen

Contractor : Jordbrugsteknisk Institut
Royal Veterinary and Agricultural University

Address : Jordbrugsteknisk Institut
Royal Veterinary and Agricultural University
Rolighedsvej 23
DK-1958 Copenhagen V

Summary:

The purpose of the project is to develop a conversion system which can produce heat energy from animal manure by combined drying, combustion and regaining of heat. Predictions of the performance show, that such a system can produce low grade heat (up to 80^{0}C) from manure with a moisture content of up to approximately 82 % w.b. Further the overall thermal efficiency is predicted to be 60-65 %.

During the first year of the project period work has been carried out on investigations concerning the characteristics of manure in relation to drying and on development of a drying unit. This work has resulted in an unit which can dry and structurize manure into a granula suitable for combustion in most types of furnaces. The consumption of electric energy for the process corresponds to about 3 % of the energy content of the dried manure.

1. Introduction.

The system which is to be developed in the project is shown schematically in Fig. 1. The function is as follows:

Wet waste is led to a drier by a feeding mechanism. When drying has taken place it passes on to combustion in a furnace. The flue gas produced by combustion is mixed up with a suitable amount of cooling air and used for drying. The humid air from the drier is led to a gas scrubber which works as combined heat exchanger and gas cleaner i.e., heat energy (incl. latent heat) and impurities are transmitted to the washing water while the cooled and cleaned flue gas is led to the open.

The water heated in the scrubber is led to a heat exchanger connected to a space heating system or similar. After cooling here it passes on to a clarification container where impurities are settled before the water is re-circulated to the scrubber. The condensate is led away through an overflow opening in the clarification container.

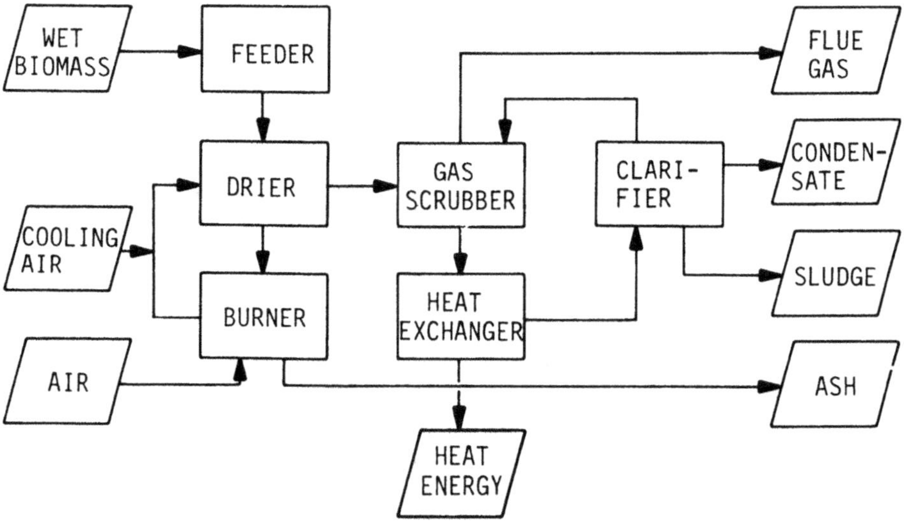

Fig. 1. Flow chart showing system for heat extraction from wet biomass by combustion. (Patent pending.)

The performance of the system has been predicted (1) to the following approximate figures:

Overall thermal efficiency ....: 60-65 %

Max. acceptabel moisture content of manure: 82 % w.b.

Max. temperature of produced heat ....: 80°C

Development work concerning drying has been in progress since medio 1980 and several experimental driers have been designed, constructed and tested in cooperation with Vølund Miljøteknik A/S, which is a manufacturing company in the field of refuse burning equipment.

The paper presents the result obtained during these experiments.

2. Apparatus and methodology.

The investigations were initiated with preliminary studies of different manure samples in order to find the characteristics at various moisture contents. Small samples of manure were dried in a laboratory oven to different moisture content and then studied in respect of consistency, strength and tendency to adhere to a steel plate.

Later on the data obtained from these studies was used in the design and forming of an experimental drier. This unit was supplied with hot air from an 11 kW electric heater with variable power output. This unit could produce air with temperatures up to approximately 700°C.

The manure was fed into the drier by a special feeding mechanism. This mechanism was designed to comminute, meter and structurize the manure into aggregates.

3. Results.

Solid manure from pigs, cattle and chickens was investigated and later on dried in the experimental drier. The following results were obtained.

3.1 Characterization of animal manure.

Solid manure from cattle and pigs typically has a moisture content of 70 to 85 %. The content of straw may vary from 10 to 50 % of the dry matter. Chicken manure has a moisture content mostly above 50 %.

The manure was found to be very absorbent and the moisture content could not be lowered to less than about 80 % by drainage. When cattle and pig manure has a moisture around the usual 80 % it is soft, although solid enough to be stacked. It adhere to nearly all materials it gets in touch with.

When drying up the strength and hardness of this manure increases gradually and it becomes a rather strong and hard material which can resist relatively rough handling. If manure dries up while in contact with another material it adhere strongly to it and is difficult to remove. During drying the manure almost maintains its outer shape and size and the density falls to about 1/4 of the initial density.

3.2. Structurizing and drying.

The processing of manure through a conversion system as the one in question is simpler and faster if the material is granulated, i.e. splitted up into small aggregates before fed into the drier. Therefore various methods of granulation was considered.

On basis of the characteristics of manure outlined in the previous section it was natural at first to try to granulate the manure just by feeding the drier with aggregates of the desired size, and to get them stabilized during drying. Several designs were tested as it proved difficult to keep the manure aggregates free of the drier surfaces until the moisture content was brought sufficiently down to avoid adherence. In the end however, a method was found which seems to work satisfactory with all types of manure with until 80 % moisture.

A typical set of main data from one of the tests is shown in table I. In relation to the drying experiments, energy consumption of the drying unit was measured to about 0.5 MJ/kg dry manure. This amount corresponds to about 3 % of the energy content of the dried manure.

Table I. Example of data from drying test with pig manure with an assumed calorific value of 18 MJ/kg dry matter.

Pig manure with approximately 20 % straw.	
Moisture content of manure	80 % w.b.
Manure flow, wet weight	2.5 g/s
Manure flow, dry weight	0.5 g/s
Flow of chemical fixed energy in manure	9.0 kW
Airflow to drier	11 g/s
Drying air temperature	545 ^{o}C
Electric power to air heater	6.5 kW
Drying capacity	2.0 g water/s
Nett power required for drying	5 kW

3.3. Characteristics of dried manure.

The developed drying system produces granulated dry manure consisting of aggregates in the size 4 to 7 mm. The amount of dust is below 5 % and the material is easy to handle with conventional conveyers and also well suited for combustion in various types of furnaces.

4. Analysis of results.

The obtained results (table I) show that the flow of energy fixed in manure exceeds the power required for drying by approximately 38 %. A manure burner connected to the drier and operated with a thermal efficiency of about 73 % therefore would be able to supply the necessary heat for drying.

On basis of the electric energy consumption measured during the drying tests, a preliminary balance can be estimated as follows:

Assuming that 1 kg of manure contains 18 MJ of energy, roughly 11 MJ is produced at the predicted efficiency of 60 % (1). In the drying unit about 0.5 MJ of electric energy is used for processing. Another 0.1 MJ may be required for water pumping to the air scrubber, for a stoker auger etc. Thus about 0.6 MJ is used for each kg of manure produced.

Another factor that must be considered is the loss of nitrogen. One kg of manure dry matter contains about 27 g N, but only about 7 g can be utilized for plant production in practice (2). The conversion system provides a possibility for regaining NH_3-N by means of the air scrubber (1) but if it is assumed that all nitrogen is wasted, the loss of energy is about 0.6 MJ/kg dry manure (84 kJ/g N according to (3)).

Thus a total of about 1.2 MJ may be required to power the conversion unit and to replace lost nitrogen. The energy ratio therefore becomes:

$$\frac{11}{1.2} = 9.2$$

when ignoring energy used for production of the conversion unit.

Conclusion.

The paper describes the results from work on development of a drier which is to be used in connection with a furnace and a heat recovering system for production of heat energy from animal manure. The main results acheived are:

1. A drying unit has been developed which can change manure with up to 80 % moisture into a granular material of dry manure aggregates.
2. The heat requirement of the drier is only about 73 % of the energy content of the dried manure. Therefore combustion of the manure can supply sufficient heat for drying.
3. The consumption of electric power for drying (structurizing and blowing) is measured to be equivalent to approximately 3 % of the energy content of dried manure.
4. Loss of nitrogen from the complete conversion system is predicted to be around 7 g N pr. kg dry matter in the worst case. The energy equivalent of this corresponds to about 3 % of the energy content in the manure.

References.

(1) Have, H. (1980):
Extraction of Heat Energy from Animal Waste by Combined Drying, Combustion and Water Vapour Condenzation. - Paper presented at the 1st Coordination Meeting of Contractors, 18/19 Sep. 1980, Amsterdam.

(2) Landbrugets Informationskontor (1978):
Håndbog for plantedyrkning.

(3) Leach, G. (1975):
Energy and Food Production. - International Institute for Environment and Development. IPC science and technology press Ltd., Guildford, Surrey.

"HIGH TEMPERATURE STRAW GRANULATION"

Contract number	ESE/071/F
Total budget	270 000 UC - CEC 30 %
Head of project	J. LUCAS C.N.E.E.M.A. Parc de Tourvoie 92160 ANTONY - FRANCE
Subcontractors	- ALSTHOM ATLANTIQUE - PROMILL

Summary

Classical granulation of straw requires glue and a lot of energy. So it is very expensive.

Tests which were carried out have shown that high temperature granulation is cheaper (no glue, less energy).

A new feeding device was placed. It will allow output up to 1 T/H.

A pellet cooler ought to be placed.

THE TECHNOLOGY OF THE "OVER-DRIER"

THE "OVER-DRIER" INSURES A HOMOGENEOUS "OVERDRYING" OF STRAW.

IT CONSIST OF :

- A CONTINUOUS AUTOMATIC FEEDING DEVICE OF CHOPPED STRAW. THE OUTPUT MUST BE PRECISELY CONTROLLED.
- A CONTINUOUS DEVICE FOR CONVEYING OUT THE OVERDRIED STRAW.
- A HEATING DEVICE OF THE UNIT.

THE EXTRUSION PROCESS

IT IS CLASSICAL (RAM PRESS OR DIE PRESS).

CONVEYING OUT THE PELLETS

THEY MUST BE COOLED.

THE WORK CARRIED OUT

- THE FEEDING DEVICE

A CONTROLLED APRON FEEDING DEVICE WAS PLACED

THE OUTPUT IS CONTROLLED BY THE SPEED OF THE APRON AND BY THE THICKNESS OF THE HOPPED STRAW ON THE APRON. THE DEVICE WAS SATISFACTORY UP TO 2 T/H OF CHOPPED STRAW.

- THE "OVERDRIER"

COMPLEMENTARY TESTS WERE CARRIED OUT WITH AN OUTPUT OF 400 Kg/H OF CHOPPED STRAW.

- CONVEYING OUT OF THE PELLETS

A PELLET COOLER OUGHT TO BE PLACED WHICH WAS NOT DONE.
THE PELLETS WERE COOLED BY SOLID CARTON DIOXID SPRAYING.

WITH A NEW FEEDING DEVICE TESTS ARE GOING TO BE CARRIED OUT FOR AN OUTPUT UP TO 1 T/H.

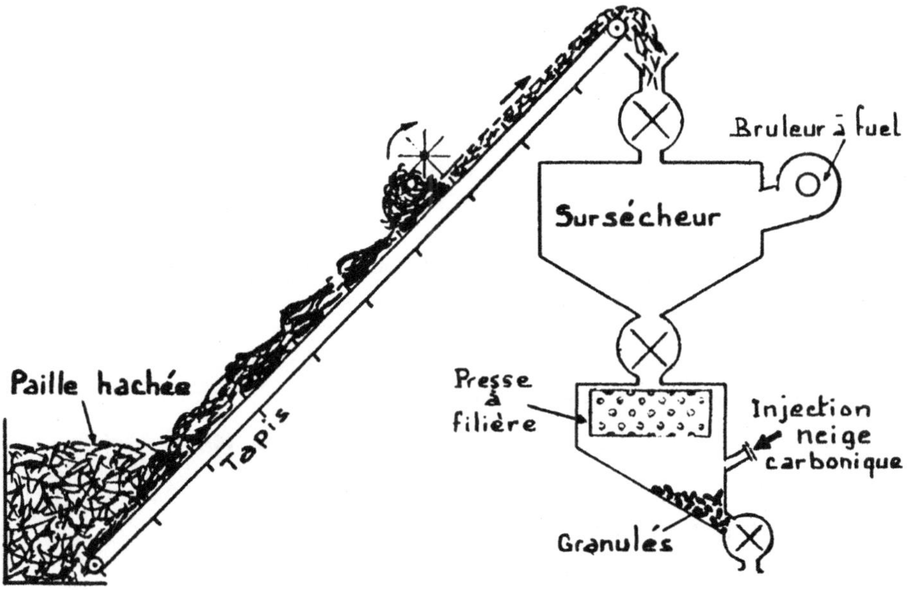
Bruleur à fuel
Sursécheur
Paille hachée
Tapis
Presse à filière
Injection neige carbonique
Granulés

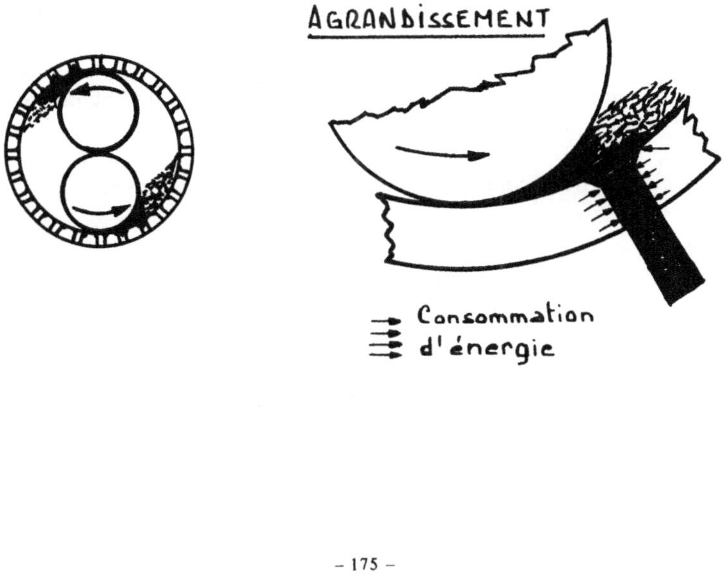
AGRANDISSEMENT
Consommation d'énergie

MOBILE PYROLISIS PLANT FOR THE PRODUCTION OF CHARCOAL AND CONDENSABLE HYDROCARBONS FROM BIOMASS.

Contract Number	: ESE - R - 048 - D
Duration	: 12 months June 1981 - June 1982
Total Budget	: 720.000 DM
Head of Project	: Johannes Linneborn, Dipl. Ing. Jan Tremus, Dipl. Ing.
Contractor	: Fritz Werner Industrie-Ausrüstungen GmbH.
Address	: D 6222 Geisenheim, Industriestraße F. R. of Germany

Summary

In order to minimize transport costs in the conversion of biomass into energy and to raise the economy of this transformation a mobile pyrolisis plant is promoted.

The installation will be located on trailers which will be moved into the production areas of biomass to enable the processing in proximity of the harvesting.

The plant will produce charcoal and liquid hydrocarbons.

1. Introduction

The different varieties of biomass are not produced in a concentrated way but on the contrary are grown on widespread surfaces. The conversion of that biomass into energy therefore calls for extensive transport to remote central plants.

It is furthermore evident, that roads are more or less rare or of unsufficient quality in these areas where biomass is produced and harvested. One should also note, that biomass contains often more than 50% of moisture. This fact signifies, that when fresh biomass is transported half of the costly moved volume is just water.

The reflections made above were the reason for the EEC to promote a project with the aim to reduce transport costs related to the transformation of biomass into energy by introducing a small mobile conversion unit instead of bigger central plants.

2. Outline of the project

The contract asks for the design and construction of a mobile plant able to produce charcoal and liquid hydrocarbons from all sorts of fresh biomass in the harvesting area.

The target of this undertaking is:
i. minimize transport effort and cost of fresh biomass
i.i. produce in situ a concentrated product of high energy content
i.i.i. raise the economy of biomass conversion.

As the contract has just started we are not able to present results for the time being and therefore have to limit us here to the description of the process and the plant which we intend to realize.

The mobile plant will be designed in such a way that it can be located on two trailers. The first trailer houses the pyrolisis plant and the electrical power supply. The second trailer is designed to support the fractional destillation, the condenser for the hydrocarbons and the storage tank for the liquid pyrolisis oil.

In this project the pyrolisis plant will be realised.

The design of the trailers considers that they have to be pulled on poor roads.

The envisaged process may be roughly described as follows:

The freshly harvested biomass will be conditioned to a size of about 20 mm.Ø. This can be done already by the harvesting machine (as foreseen in the EEC sponsored project with "CEMAGREF") or by means of a chipper provided with the two trailers.

The biomass will then be dried and introduced into the pyrolisis plant. The pyrolisis process takes place on a moving bed which is externally heated. The cooling of the produced charcoal will be performed in a second moving bed. The aim of this dynamic pyrolisis and cooling is to minimize the contact area necessary and to maximize the introduction and extraction of heat.

Several tests are actually under way to determine the best suited solutions for the beds and the heating/cooling.

The capacity of the pyrolisis plant is designed to treat 20 tons of biomass per day, approx. 300 kg per hour.

The annexed graphic sketches show the functioning of the pyrolisis plant, the destillation and the energy balance of the whole process.

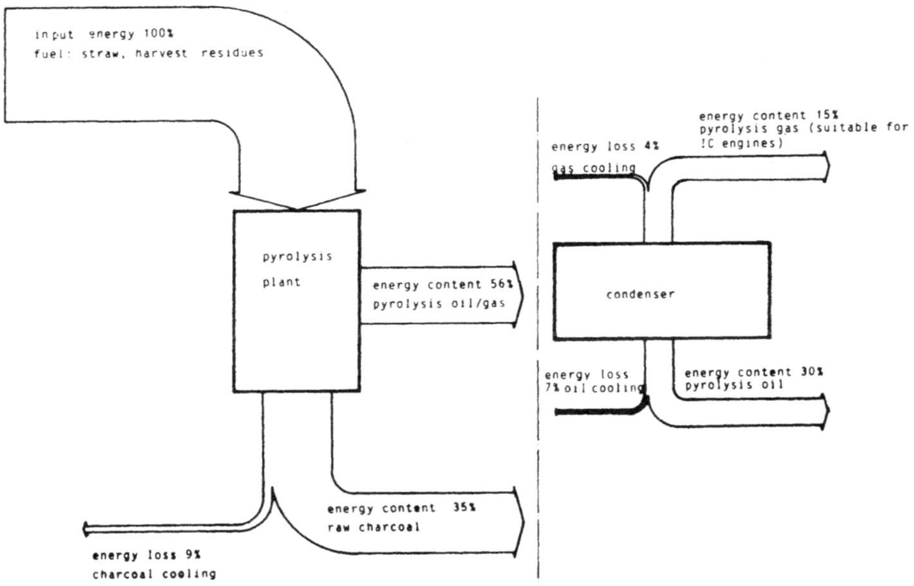

Energy Balance
for the mobile pyrolysis plant

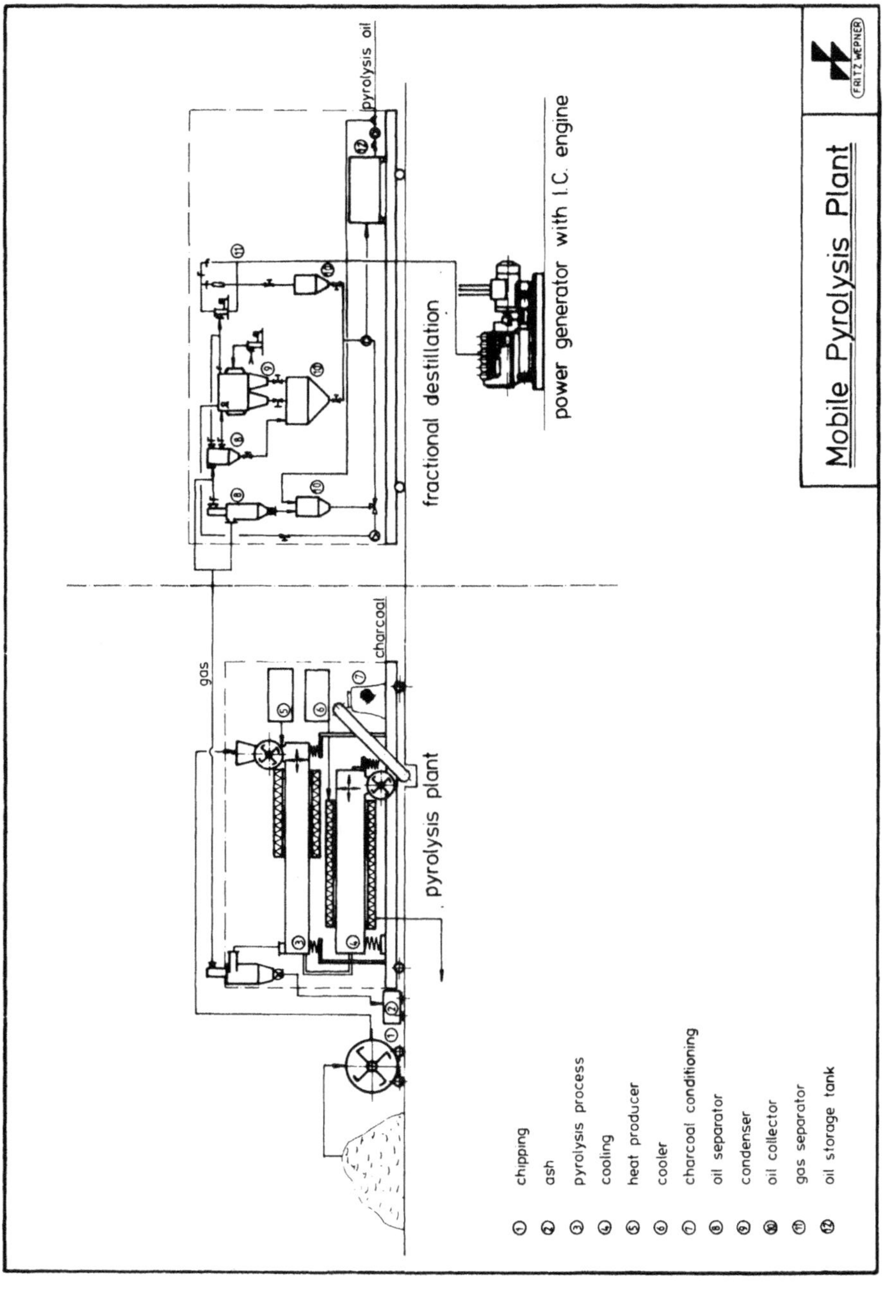
Mobile Pyrolysis Plant
FRITZ WEPNER
pyrolysis plant
fractional destillation
power generator with I.C. engine
gas
charcoal
pyrolysis oil
① chipping
② ash
③ pyrolysis process
④ cooling
⑤ heat producer
⑥ cooler
⑦ charcoal conditioning
⑧ oil separator
⑨ condenser
⑩ oil collector
⑪ gas separator
⑫ oil storage tank

THERMOCHEMICAL BIOMASS GASIFICATION

Contract Number : ESE-R-029-NL

Duration : 36 months 1 July 1980 - 30 June 1983

Total Budget : Dfl 986,000 CEC contribution : Dfl 304,000

Head of Project : Prof.dr.ir. W.P.M. van Swaaij
Dr.ir. A.A.C.M. Beenackers

Researcher : Ir. F.G. van den Aarsen

Contractor : Twente University of Technology
Department of Chemical Reaction Engineering

Address : P.O.Box 217
7500 AE ENSCHEDE
The Netherlands

Summary

The production of a clean fuel gas by means of biomass gasification in a co-current moving bed reactor has recently been commercialised. Further development of this type of reactor is concentrated on the recycle of pyrolysis gases at high specific throughputs.
Research has started on the development of a second generation gasification process, consisting of a fluidised bed gasifier in combination with a heat exchanger-tar convertor. Several bench scale units and a pilot plant fluid bed gasifier have been constructed.
Preliminar results on the flash pyrolysis of beech wood are presented.

Introduction

The process of thermochemical biomass gasification in a co-current moving bed gasifier has been developed in our department under EEC contract nr. 662-78-1ESN.
The outcome of this development has been reported (1-8) and meanwhile the process has been commercialized by a Dutch company "Machinefabriek Oldenzaal". The co-current moving bed gasifier has the advantage of a simple construction, high carbon conversion little treatment of the feedstock and low tar content of the product gas. The gas can be used in modern power and heat generating equipment.
However some problem areas still exist being:

- gas tar content at high capacities
- small sized feestock difficult to process
- hight ash content feedstock difficult to process.

Above mentioned disadvantages have led to a follow up in research and development under the present contract.

Research and Development Programme

To counteract the problem of high tar content in the product gas of a co-current moving bed gasifier at high capacities, recycling of pyrolysis gases to the feed air followed by combustion is researched. Two types of pyrolysis gas circulating systems have been constructed and the systems now are tested in two biomass gasifiers.
We hope to report on the results in near future.

Next to this the development of a second generation thermochemical gasification process has started. The aim of this development is to get a process that:

- can accept a wide variety of feedstock (including small sized and ash rich fuels)
- has a low tar content in the product gas at high throughputs
- has a high overall thermal efficiency.

We expect these advantages from fluidised bed gasification followed by tar cracking in an efficient heat exchanger, resulting in preheated gasifying agent (in this case air) and a clean product gas. Fig. 1 indicates the basic concept of this process.
The present research programme subjected to this process is along the following lines:

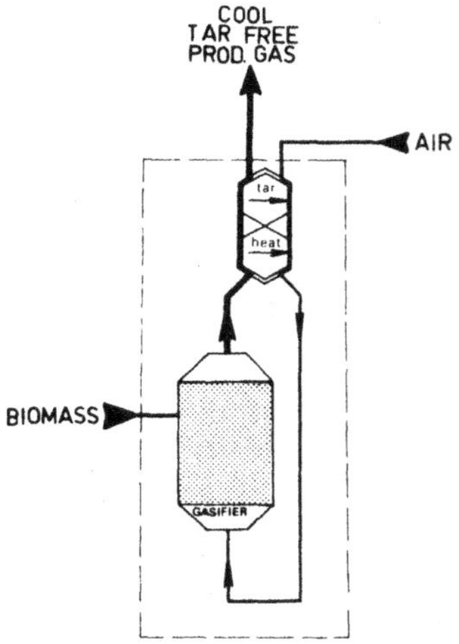

Fig.1. General gasification process scheme

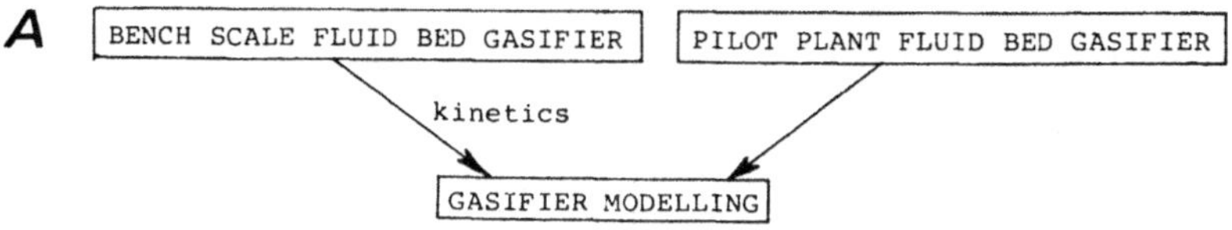

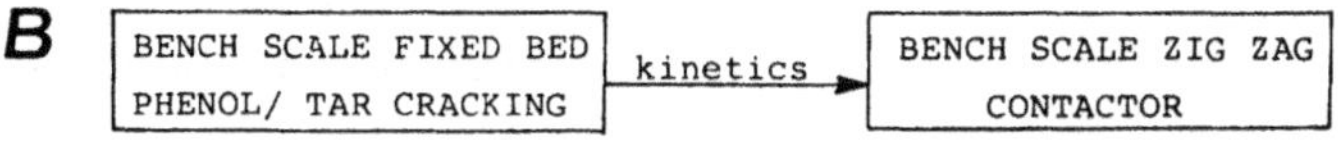

Bench scale fluidised bed reactor.

Both the yields of the flash pyrolysis proces and the char gasification kinetics are studied on a bench scale. A 30 mm diameter quartz reactor has been constructed (fig. 2) which can be continuously fed with beech wood particles (1-2 mm in diameter). During pyrolysis experiments N_2 is the fluidising medium and in gasification experiments a mixture of nitrogen, water and CO_2 is applied. The reactor temperature can be varied up to 1000 °C.

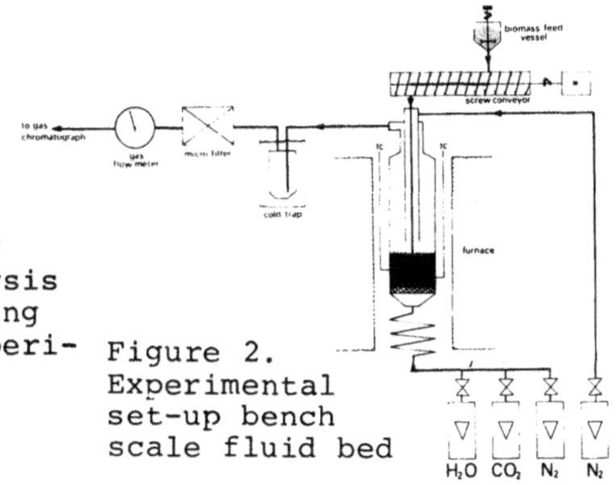

Figure 2. Experimental set-up bench scale fluid bed

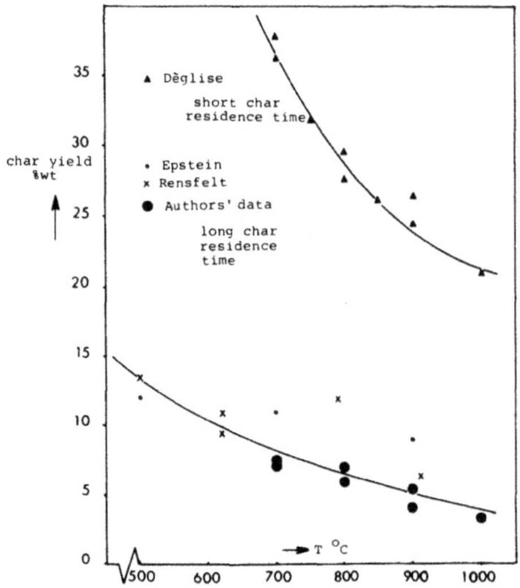

Char yield in flash pyrolysis processes as a function of temperature

Figure 3.

Preliminar flash pyrolysis results at residence times of the char typical for the fluidized process (15-20 min) show the char yield as a function of pyrolysis temperature (fig. 3).

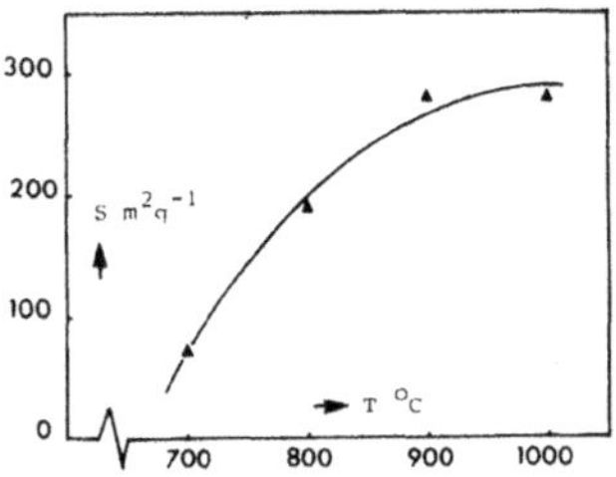

B.E.T. surface area of char from flash pyrolysed beech wood

Figure 4.

These results are in good agreement with similar results from open literature (Epstein (9) and Rensfelt (10)). For comparison fig. 3 gives also literature results at residence times of a few seconds (Déglise (11)).
Table I gives the char elemental analysis as a function of pyrolysis temperature; the influence of residence time is given as well.

Table I.
Elemental composition of char from flash pyrolysed beech wood

T °C	Author's data (long char res. time)	Dèglise (short char res. time)
beech wood	$CH_{1.48}O_{0.69}$	$CH_{1.44}O_{0.66}$
700	$CH_{0.38}O_{0.11}$	$CH_{0.90}O_{0.24}$
800	$CH_{0.23}O_{0.05}$	$CH_{0.73}O_{0.19}$
900	$CH_{0.05}O_{0.08}$	$CH_{0.58}O_{0.14}$
1000	$CH_{0.06}O_{0.13}$	$CH_{0.43}O_{0.07}$

As we want to use the flash pyrolysis char for kinetic studies, the BET surface of the fifferent chars' has been determined (fig. 4); a sharp increase of BET surface with increasing pyrolysis temperature is shown; which also indicates that a sharp rise in gasification reactivity can be expected.

Pilot plant fluid bed gasifier

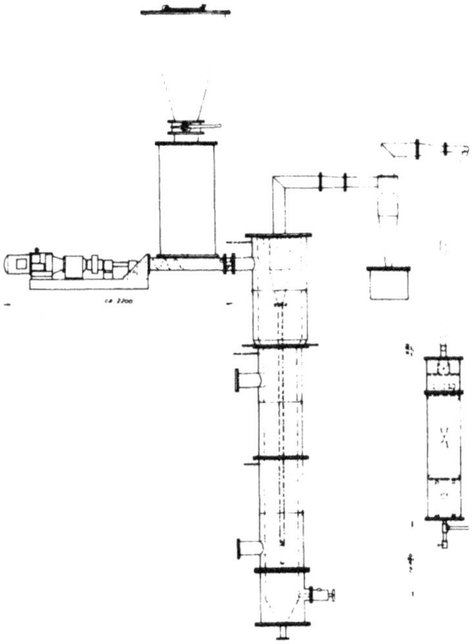

Figure 5.
Pilot plant fluid bed gasifier

A 25-50 kg wood/hr fluid bed gasifier (fig. 5) is being erected; the pilot plant is planned to start-up early July. From pilot plant operation, information on thermal efficiency, gas quality, flexibility etc. will be extracted. Next to this it will provide us with data about mass tranfer, mixing effects and overall fluid bed behaviour. Together with the bench scale reactor results this will enable us to build a reactor model, which is required for further scaling up of the process.

Bench scale fixed bed reactor

A 10 mm diameter fixed bed reactor was constructed (fig. 6) in which homogeneous and heterogeneous cracking kinetics can be studied. Phenol is used as a modelling compound. Preliminar experiments look very pro-

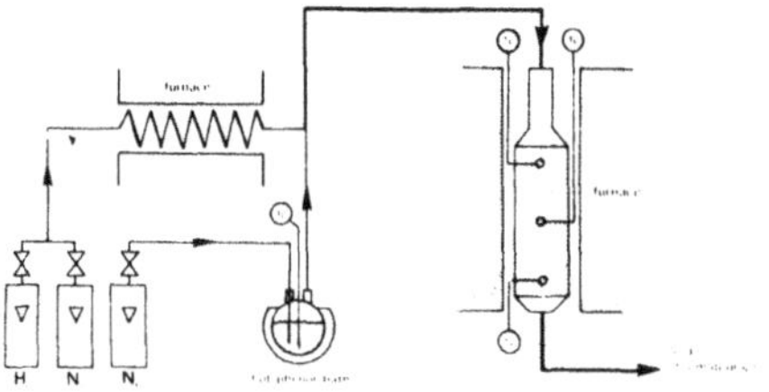

Figure 6. Experimental set-up bench scale fixed bed

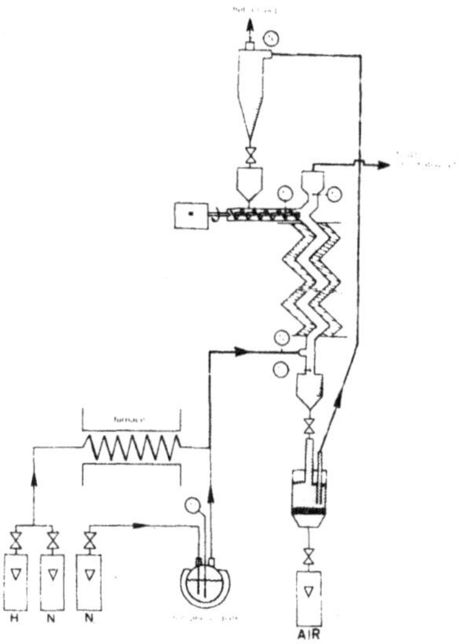

Figure 7. Experimental set-up bench scale zig zag contactor

mising and we plan to report on this in detail next year.

Bench scale zig zag contactor

A 20 mm diameter zig zag contactor (fig. 7) in which solids and gas can be contacted countercurrently. This type of contactor has been developed recently in our laboratory as a heat exchanger (12-20) and it is our aim to operate this apparatus as a combined heat exchanger -tar cracker. Data from these experiments together with kinetic data obtained from the fixed bed reactor experiments will enable us to describe the cracking process in the heat exchanger.

Coming year activities.

As last year was mainly dedicated to design and construct experimental set-ups and a fluidised bed pilot fluid bed gasifier, coming year will be one of gathering experimental results on:

- the effects of recyling pyrolysis gases in a co-current moving bed gasifier
- yield of flash pyrolysis in a bench scale fluid bed
- gasification kinetics of flash pyrolysed char
- behaviour of a pilot plant fluid bed gasifier
- cracking kinetics of phenol and/or wood tar
- behaviour of a bench scale zig zag contactor.

Literature

1. Groeneveld, M.J. van Van Swaaij, W.P.M., Gasification of solid waste, Aplied Energy 1979, 5, 165-179.
2. Groeneveld, M.J. and Van Swaaij, W.P.M., The design of co-current moving bed gasifiers fueled by biomass, ACS symposium September 1979, Washington D.C.
3. Groeneveld, M.J. and Westerterp, K.R., Social and economical aspects of the introduction of gasification technology in the rural areas of developing countries e.g. Tanzania, ibid.
4. Hos, J.J.; Koppen de Neve, P.; Groeneveld, M.J. and Van Swaaij, W.P.M., Co-current gasification of solid wastes,

Int. Recycling Congress Berlin 1979.
5. Groeneveld, M.J. and Van Swaaij, W.P.M. Gasification of char particles with CO_2 and H_2O, 6th International Symposium of Chemical reaction Engineering Nice, France, March 1980.
6. Groeneveld, M.J.; Ph.D. Thesis, Twente University of Technology (1980).
7. Hos, J.J.; Groeneveld, M.J. and Van Swaaij, W.P.M., Gasification of organic solid wastes in co-current moving bed gasifiers, IGT Symposium Orlando, Florida, 1980.
8. Final report of Contract 662-78-IESN between EEC and Twente University of Technology, Enschede (1981).
9. Epstein, E. et.al.; I.G.T. Symp. "Energy from biomass and wastes" Aug. 1978, Washington D.C.
10. Rensfelt, E. et.al.; I.G.T. Symp. "Energy from biomass and Wastes" Aug. 1978, Washington D.C.
11. Déglise, X. et.al.; Conf. "Energy from Biomass", Brighton Nov. 1980.
12. A.W.M. Roes; W.P.M. van Swaaij, Hydrodynamic behaviour of a gas-solid packed column at trickle flow, Chem. Eng. J. 1979, 17, 81-89.
13. A.W.M. Roes and W.P.M. van Swaaij, Axial dispersion of gas and solid phases in a gas-solid packed column at trickle flow, Chem. Eng. J. 1979, 18, 29-37.
14. A.W.M. Roes and W.P.M. van Swaaij, Mass transfer in a gas-solid packed column at trickle flow, Chem. Eng. J. 1979, 18, 29-37.
15. A.W.M. Roes and W.P.M. van Swaaij, An evaluation of radial solid spread factors in a gas-solid packed column at trickle flow, Chem. Eng. Sci. 1979, 34, 133-135.
16. A.W.M. Roes and W.P.A. van Swaaij, Straaltjes stroming (trickle flow) van een vaste stof en een gas in een gepakte kolom.
17. A.W.M. Roes and W.P.M. van Swaaij, Gas/Feststof Rieselströmung in einer Füllkörperkolomm, Chem. Ing. Tek. 1979, 51, Synopse 694.
18. I.W. Noordergraaf; A.W.M. Roes and W.P.M. van Swaaij, Experimental study on the behaviour of a gas-solid counter-current zigzag contactor at trickle flow, Proc. 2nd Powder Europ. Conf., Wiesbaden, I 80.
19. I.W. Noordergraaf; A.W.M. Roes and W.P.M. van Swaaij, Axial mixing and mass transfer in a zigzag contactor, Proc. 3rd Int. Symp. on Fluidization, Henniker, U.S.A. VIII 80.
20. I.W. Noordergraaf; A.B. Verver and W.P.M. van Swaaij, Gas/Feststoff Rieselströmung und Austausch in einem Zickzack-kontaktor, TIZ Fachber. 104, 1980, 534-6

PRESSURISED GASIFICATION OF WOOD WITH OXYGEN COMBUSTION OR ELECTRICAL HEATING

Contracts Number	:	ESE-023-F
Duration	:	12 months 1 October 1980 - 1 October 1981
Total Budget	:	925 000 FF CEC Contribution : 40 000 E.U.A. (∿ 160 000 E.U.A.)
Head of Project	:	Pierre DUBOIS Laboratoires de Marcoussis, Centre de Recherches de la CGE Route de Nozay - 91460 MARCOUSSIS (France)
Contractor	:	NOVELERG
Address	:	12 rue de la Baume 75008 PARIS
Subcontractors	:	. Laboratoires de Marcoussis (Centre de Recherches de la CGE) . Ateliers et Chantiers de Bretagne . CGEE-Alsthom

Summary

The general purpose of the Compagnie Generale d'Electricité (C.G.E.) program is to be able to design and build plants for methanol production from wood in Europe or in other parts of the world. The C.G.E. program is conducted by Novelerg and carried out by other CGE subsidiaries :

- Les Laboratoires de Marcoussis, the CGE Research Center,
- Les Ateliers et Chantiers de Bretagne,
- CGEE-Alsthom.

The general study of a 50 tons a day methanol plant is now in progress in close cooperation with Technip. The following tasks on the experimental study of wood gasification have been completed :

- we have designed and built a laboratory apparatus which has been used for the last 6 month for tests which show the strong influence of pressure on the kinectics of pyrolisis and gasification,
- a test gasifier of significant size (height of 13,5 meters, diameter of 1,3 meter) has been designed and is now at the final stage of manufacturing by Ateliers et Chantiers de Bretagne in Nantes ; this gasifier will be sent to Marcoussis in July for extensive tests of both the key components such as the pressurized lock hopper for wood feeding and the complete system : influence of pressure, influence of the mass flow of recycled gases, comparison of electrical heating to the oxygen combustion of wood to provide the high temperature heat needed for wood gasification.

1. Introduction

The general purpose of the Compagnie Generale d'Electricité (C.G.E.) program is to be able to design and build plants for methanol production from wood in Europe or in other parts of the world.

Most of the research and development work is done on wood gasification under pressure as this is the only really new part in the methanol synthesis from wood. The C.G.E. program is conducted by NOVELERG and carried out by other CGE subsidiaries :

- Les Laboratoires de Marcoussis, the C.G.E. Research Center,
- Les Ateliers et Chantiers de Bretagne,
- C.G.E.E.-Alsthom.

For the part of the plant producing methanol from the product gases, Novelerg works in close cooperation with Technip, which has a large experience in the design of plants producing methanol from natural gas.

2. 1980-1981 program

In 1980-1981, the following tasks will be completed :

- design study of a plant which would have a methanol production capacity of 50 tons/day,
- experimental study of wood gasification under pressure :
  - . pyrolysis and gasification kinetics studies with a laboratory apparatus,
  - . design, manufacturing and test of a high pressure gasifier, converting two tons of wood per day at the atmospheric pressure.

On this gasifier, we will compare energy balances, amount and quality of the gas produced for several values of the mass flow of recycled gases and for two ways of energy input for the endothermic reactions :

- combustion of wood or pyrolineous gas with oxygen,
- electrical heating and then there is no need for oxygen and we have a "steam only" gasification.

3. Electrical heating compared to wood burning with oxygen

Electrical heating looks promising, from an economic point of view, when electricity is available at low cost as it happens during off peak hours in countries with important nuclear programs or, most of the time in countries with large hydroelectrical power production like Brazil, Canada or Africa.

For the production of 1 kilo of methanol from wood containing 30 % of water in weight, one needs :

- either 2,3 kg of wood and 0,4 m^3 of oxygen,
- or 1,6 kg of wood and 2,4 kWh of electricity.

So, by spending 2,4 kWh of electricity, one can save 0,7 kg of wood and 0,4 m^3 of oxygen. For a price of wood ranging from 2 U.S. cents in America or in Africa to 4 U.S. cents in Europe, the break-even point in raw materials costs corresponds to a price of electricity of 2 cents per kWh in Europe and 1,4 cents in America or in Africa. The economic interest of electrical heating derives from the fact that electricity produced from hydroelectrical generators costs today one cent in Canada or in Brazil and that in 1985 electricity will be available in France to industry at a price ranging from one to two U.S. cents for more than 4 000 hours a year.

This very simple cost evaluation show that electrical heating with no need for oxygen input instead of wood combustion with oxygen can be a simple and cost effective way to bring in the gasifier the high temperature heat which is needed for wood gasification.

4. Work progress

The general design study of a 50 tons a day methanol plant is now in progress in close cooperation with Technip. As far as experimental studies of wood gasification are concerned, the following tasks have been completed :

- we have designed and built a laboratory apparatus which has been used for the last 6 month to study the kinectics of pyrolisis and gasification under pressure,
- a test gasifier of significant size has been designed and is now at the final stage of manufacturing by Ateliers et Chantiers de Bretagne in Nantes ; this gasifier will be sent to Marcoussis in July for extensive tests.

4.1. Laboratory apparatus for pyrolisis and gasification studies

We have designed and built a laboratory apparatus for pyrolisis and gasification kinectics studies. A pressurized vessel (cf. figure 1) with oxygen and steam inlets, with internal electrical heating and good thermal insulation can contain one liter of wood or charcoal. This vessel can be operated up to 20 bars and 900°C. The amount of H, CO, CO_2 and CH_4 contained in the gas produced is measured by a chromatograph.

Interesting results have been obtained showing a strong influence of pressure on the pyrolisis and gasification kinetics. For the pyrolisis at 300°C of 8 cm^3 oak cubes, one can see on figure 2 that when pressure goes up from one to 6 bars, the time needed to produce 90 % of the gas, which goes down from 38 minutes to 16 minutes, varies very nearly like the square root of pressure. And, in this experiment, starting from room temperature, the time before any significant pyrolisis happens, went down from about 45 minutes to about 12 minutes.

4.2. Design and construction of a test gasifier of significant size

We have designed these gasifier of significant size shown on figure 3. The main pressure vessel has a 1,3 meter external diameter and a 13,5 meter height. It is a fixed bed downdraft gasifier which can be operated up to 20 bars.

On the recycled gases loop, a 150 kW electrical heater or a combustion chamber with steam and oxygen inlets can be used to bring in the gasifier the high temperature heat needed for wood gasification. The blower on the recycled gases loop will be driven by a variable speed electrical motor so that we can study the influence of recycled gases mass flow on the gasifier gas production.

Special care must be taken of safety problems both at the design stage and during the tests of gasifier, due to the combination of high pressure and high temperature operation and toxic and flammable gases. Special attention has been payed to wood feeding and ashes removal and to the combustion chamber.

Wood will be fed on a batch basis through a pressurised lock hopper (see figure 4). The lock interspaces will be alternatly pressurised with gas produced by the gasifier or filled with air after the gas has been sent to the flare. A vacuum pump will prevent any mixture of air and gas during the lock hopper operation.

The gasifier, now at the final stage of manufacturing by Ateliers et Chantiers de Bretagne will be sent to Marcoussis in July for extensive tests of both the complete system and the key components.

Figure . 1

VESSEL FOR PYROLYSIS AND GASIFICATION KENETICS STUDIES

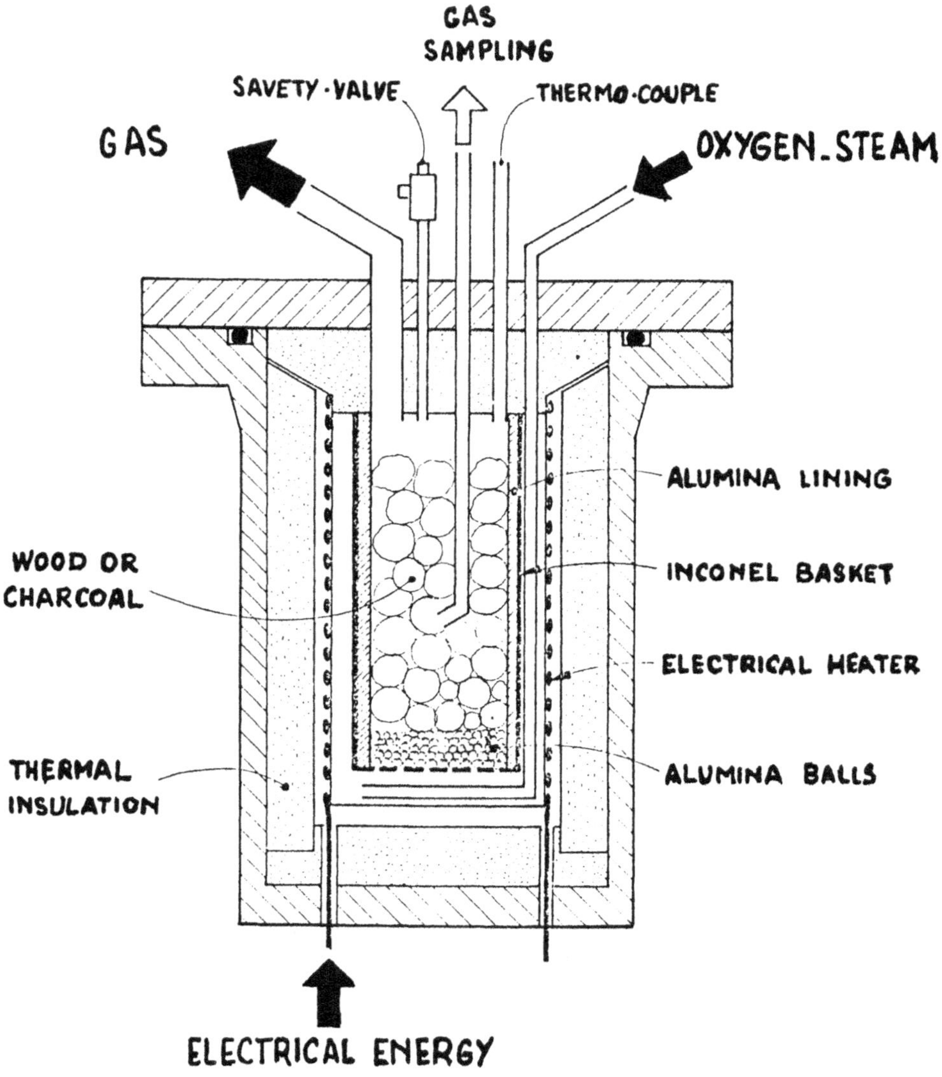

Figure .2 INFLUENCE OF PRESSURE ON THE PYROLYSIS AT 300°C OF 8 cm^3 OAK CUBES.

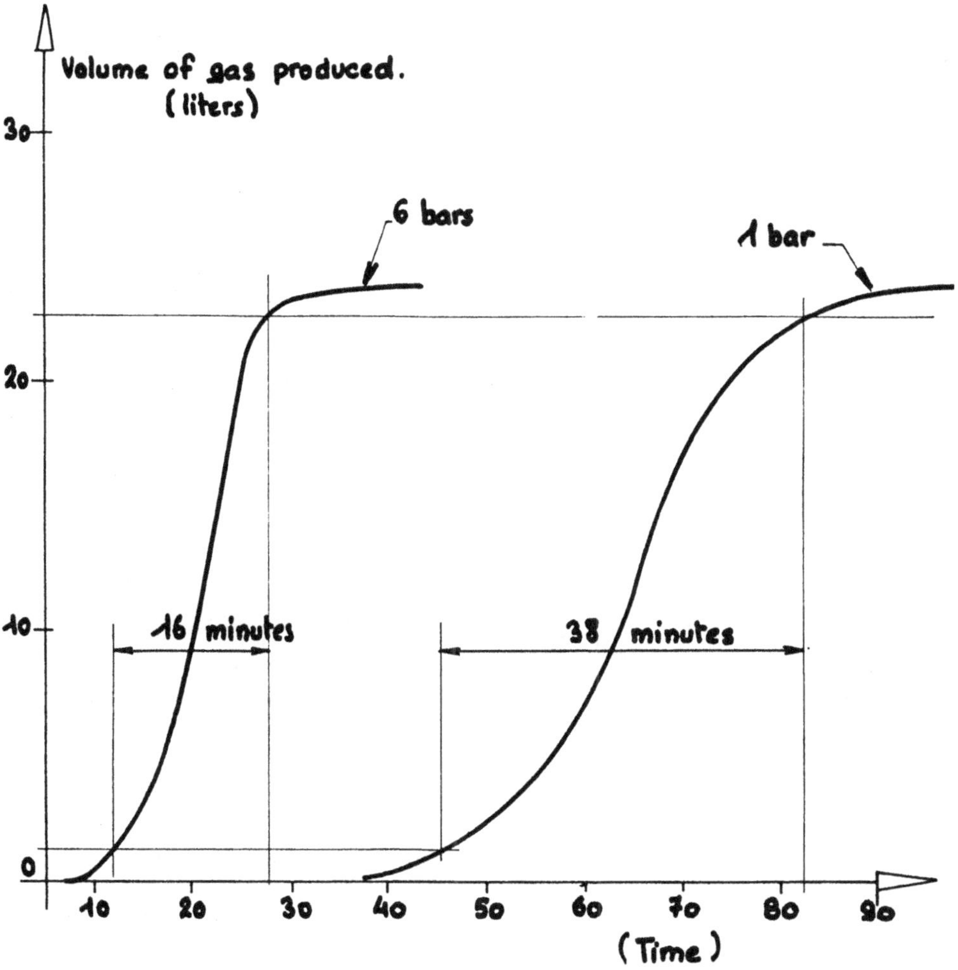

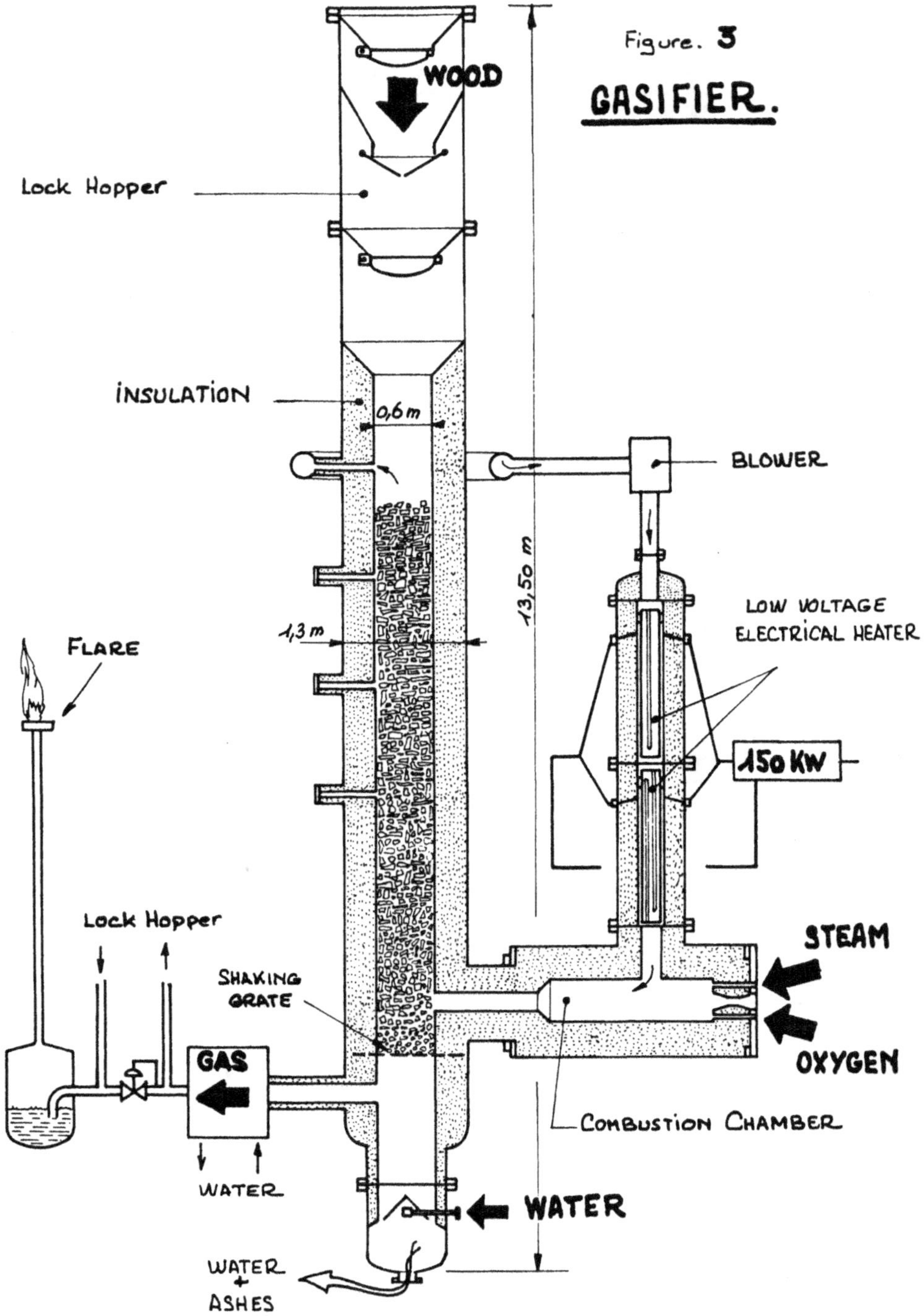
Figure. 3
GASIFIER.
WOOD
Lock Hopper
INSULATION
0,6 m
BLOWER
13,50 m
LOW VOLTAGE
ELECTRICAL HEATER
1,3 m
FLARE
150 KW
Lock Hopper
STEAM
SHAKING
GRATE
GAS
OXYGEN
COMBUSTION CHAMBER
WATER
WATER
WATER
+
ASHES

Figure. 4 WOOD FEEDING

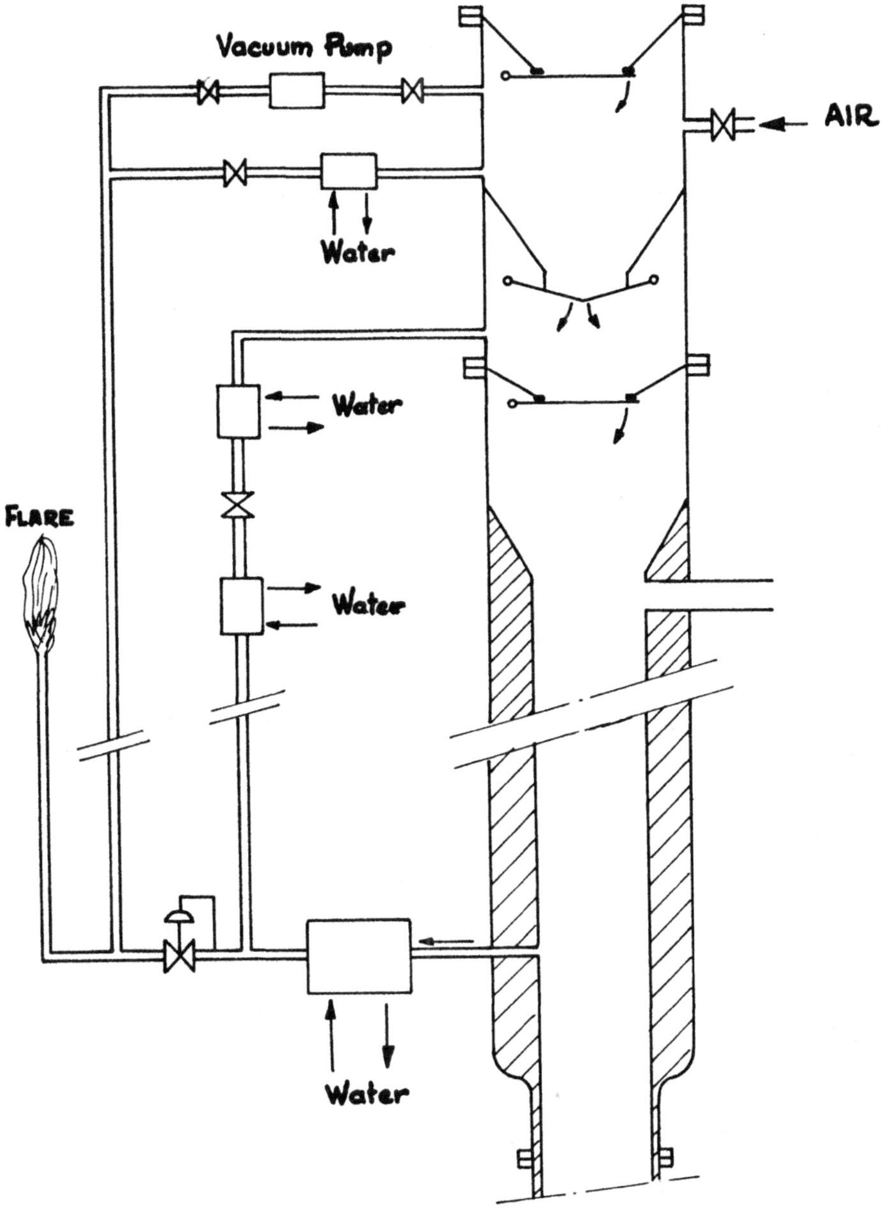

EXPERIMENTAL WORK ON A FIXED-BED OXYGEN GASIFIER IN THE VIEW OF METHANOL SYNTHESIS USING BIOMASS AS A FEEDSTOCK

Contract number : ESE - R - 030 - F (G)

Duration : 12 months from 15/11/80 thru 14/11/81

Total Budget : 1.352.000 F.F.

CEC Contribution : 608.400 F.F.

Main contractor : CREUSOT-LOIRE ENTREPRISES (CLE)
33, Quai Galliéni
92150 - SURESNES

Other contractors :
- Centre National du Machinisme Agricole, du Génie Rural, des Eaux et des Forêts (CEMAGREF)
- Centre Technique Forestier Tropical (CTFT)
- Société Nationale Elf-Aquitaine (SNEA)
- Total Energie Développement (TED)

Head of Project : Mr. Michel COURTINAT, Directeur Energies et Techniques Nouvelles, Creusot-Loire-Entreprises.

It is now well recognized that methanol will play a more extensive role as a fuel in the future than in the past. Renewed interest has risen in the manufacture of methanol from various sources, such as biomass. To convert solid carbonaceous materials to methanol, it is necessary to partially oxidize the material to form a synthesis gas consisting primarily of H_2 and CO. This can be done in a gasifier using oxygen, rather than air, to oxidize the feed material.

For several years, the CEMAGREF (formerly CNEEMA) has worked in the field of biomass gasification, developing reactors for producing low-Btu gas using air as the oxidizing medium. Based on the same concepts, an oxygen fixed-bed gasifier in the view of producing synthesis gas is being developed under this contract. Five French organizations are putting their efforts together in order to achieve a quick positive result, the leading organization being Creusot-Loire-Entreprises.

The aim of the contract was to build a gasifier having a capacity of 5 metric tons of biomass/hr, and to thoroughly test it at atmospheric pressure. Such an equipment has been constructed using parts of a previous experimental unit, the nominal capacity of which being in the range of several hundred kg of biomass/hr. One of the basic feature of this gasifier is a recycling system which allows heavy hydrocarbon compounds to be burnt and thus eliminated from the exhaust gas. Another fundamental concept is to adjust the temperature such as ashes are not melted and do not cause any blocking problems.

Lots of experiments have been conducted using small wood logs or big size wood chips as a feedstock. Main results so far can be summarized as follows :

- the system is reliable, and continuous experimentation for 36 hours with pure oxygen has been realized with stabilized conditions.

- no melting of ashes has been observed, neither during air gasification or pure oxygen gasification.

- almost no heavy hydrocarbons could be detected in the exhaust gas, and methane content is below 2 %.

- balance of products have been established within 10 %. Leakage of the equipment is the main reason for the 10 % loss. A typical composition of exhaust gas is described in the attached chart : 26 % CO and 30 % H_2 have been obtained, even if the gas still contained 14 % nitrogen.

- a feed rate of 1.25 tons/hr has been obtained using humid wood (30 % humidity content). Increase of the capacity will be investigated shortly.

Previous theoretical studies have shown the advantage of operating gasifiers at pressures above atmospheric pressure for production of methanol. Beside the development of this fixed-bed gasifier, our Group is also investigating an entrained-bed gasifier to process light biomass products at pressures in the 10 - 35 bars range.

The combined results of these development work should shortly lead us to start the detailed engineering and the construction of a pilot unit having a capacity of 5 - 15 tons of biomass/hr. Such a size appears well suited for methanol production using biomass as a feedstock.

CEMAGREF FIXED-BED GASIFIER

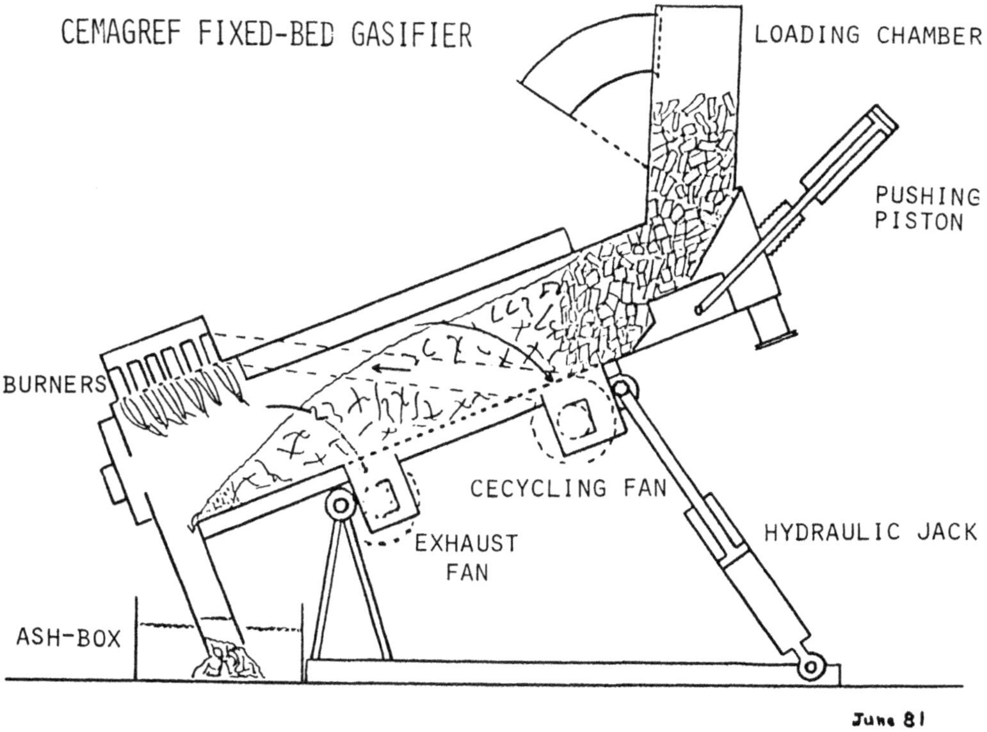

CREUSOT-LOIRE ENTREPRISES

le 24 juin 1981
JR/mlo

OXYGEN GAZEIFICATION OF WOOD

Example of mass balance

(for 1 t of wood)

I. WOOD ANALYSIS

Moisture	:	30 %
Dry wood ($C_6H_9O_4$)	:	70 %

II. GAS ANALYSIS (1 689 kcal/Nm3)

	H_2	O_2	CO	CO_2	CH_4	N_2	H_2O
Dry gas	30	0	26	28,5	1,5	14	
Wat gas	21	0	18,2	20	1,0	9,8	30

III. BALANCE

	C	O_2	H_2	N_2	TOTAL
IN					
From dry wood	347	310	43		700
wood moisture		267	33		300
air (leaks)		55		180	235
oxygen		250			250
TOTAL	347	882	76	180	1 485
OUT					
From H_2			28		28
CO	143	191			334
CO_2	157	420			577
CH_4	9		3		12
N_2				180	
H_2O		315	39		354
TOTAL	309	926	70	180	1 485
DEVIATION	- 38	+ 44	- 6		
	- 11 %	+ 5 %	- 8 %		

PROCESS AND EQUIPMENT FOR THE FLUIDIZED BED OXYGEN GASIFICATION OF WOOD

Contract number	:	ESE-R-031-F
Duration	:	12 months 1 January 1981 - 31 December 1981
Total budget	:	525 500 FF CEC Contribution : 236 000 FF
Head of project	:	G. CHRYSOSTOME
Contractor	:	CREUSOT-LOIRE
Address	:	42, rue d'Anjou 75008 PARIS
Represented by	:	M. MARCELLIN

Summary

The aim of the research is to develop fluidized bed oxygen gasification of wood. A new process as well as the equipment allowing its implementation are to be developped. The combination of fluidized bed and oxygen allows to obtain a medium BTU gaz free of any nitrogen. The use of fluidized bed allows to reduce wood preparation and feeding.
The research project involves the following stages :

- getting kinetic and thermodynamic data with the help of a low thermal inertia research reactor (Université de Technologie de Compiègne),
- developping and implementing a fluidized bed gasifier that may be fed with wood, bark or sawdust,
- operating the pilot reactor under various conditions.

At the end of June 1981, about 30 tests have been run :
- combustion of wood with air,
- gasification of wood with air,
- gasification of wood with air plus pure oxygen,
- gasification of wood with pure oxygen.

All tests being performed with an inert material as fluidized bed.

1 - Introduction

The planned research will allow the use of wood, a considerable amount of it being available in the European Community. Wood is a renewable energy source.

This research opens the way to synfuels (methanol) and will contribute to reduce oil imports. Poor fuels such as oil and coal shales could probably also be processed.

Last the process equipment can be exported to non EEC countries (for example for tropical wood applications) and thus provide currency required to purchase row materials whether energetic or not.

2 - Description of the research and of the equipment

The program consists of two main points.

2.1 - Design and implementation of the reactor (figure I)

The major piece of equipment is a fluidized bed gas generator, 400 mm internal diameter. In order to reach a steady load operation, as quick as possible, we have been using a low thermal inertia unit. Our gasifier includes a steel casing, insulated from the outside, with, as its base, a porous or a perforated plate supporting the fluidized bed. Regarding the fluidized bed material, two solutions are to be tested.

1 st : use of a fluidized bed made of carbon (charcoal) resulting from wood pyrolysis. The fluidized bed level shall however remain constant as the oxygen injected through the fluidizing plate allow the gasification of the carbon and avoid any progressive building-ups in the reactor.

2 nd : use of a fluidized bed made of inert material. Then, fluidized bed contains low quantities of carbon and wood is immediately gasified.

The first solution shall certainly allow better thermal control than the second. However, tars emissions could occur. The second solution is more adequate for gasification of wood chips of various dimensions.

The reactor is provided with a sufficient number of temperature probes to establish the unit thermal chart. A differential pressure gauge controls the pressure under the plate and above the fluid bed.

By holding the pressure in the reactor slightly below atmospheric pressure, emission of harmful gas (CO) in the station can be minimized. Air and/ or oxygen can be blown into the fluidized bed.

Part of the gas discharged by the gasifier is sent to gas analysers. The major part of the produced gas is fired in a special burner into an ancilliary combustion chamber in order to avoid any discharge of polluting gases in the atmosphere.

Finally a cyclone and a wood injection system have been added to the reactor. Wood is fed into the fluid bed by an endless screw system. If necessary sawdust can be sent through a pneumatic system within the fluidized bed itself.

We measure continuously all fluid and solid flows with appropriate devices.

2.2 - Gasifier operation

Tests planned will include :

- wood oxygen gasification on inert material fluidized bed,
- wood oxygen gasification on carbon fluidized bed.

The major parameters being tested, which affect the resulting syngas, are :

- fluidization gas composition,
- moisture of wood,

- fluidized bed temperature,
- fluidizing velocity,
- influence of the equivalence ratio : $\frac{O_2 \text{ really injected}}{O_2 \text{ stoechiometric}}$
- dimensions of the wood chips.

Besides their influence on the resulting gas analysis the effect of such parameters on the rate of elutriation and on the tar content are studied.

We take periodical samplings when the unit is operating allow to detect any possible enrichment of fluidized bed by alkaline elements able to start the sintering of this fluidized bed while in operation on a continuous industrial basis.

3 - Results and comments

We have started equipment testing at the beginning of January. Tests run, until the end of June, have been using the following :
- an inert fluidized bed material (sand containing 99 % $Si\ O_2$)
- wood chips as fuel from 2 to 40 mm (pine - tree)

The scheduling for equipment testing was :
- 1 st : combustion of wood with air,
- 2 nd : gasification with air
- 3 rd : gasification with air + O_2
- 4 th : gasification with pure O_2

Qualitatively the first results obtained indicate that :
- composition of syngas is affected priorily by moisture of wood. To obtain a gas of significant heating value it will be necessary to use a predried wood so that its moisture don't exceed 20 %.
- to obtain an important yield in the cracking of volatile compounds of wood, residence time in the reactor and temperature have to be high enough. (Temperature : 900°C ; time : 5 s).
- The equivalence ratio E.R : $\frac{O_2 \text{ injected}}{O_2 \text{ stoechiometric}}$ depends on the moisture of wood. For example, with a 30 % moisture wood, at 850°C the E.R is 0,6 and the lower heating valve of syngas 1550 kcal/Nm^3. With a 10 % moisture wood at the same temperature E.R is 0,38 and the lower heating value of sÿngas 2250 kcal/Nm^3.
- Ashes contained in wood don't stay in the fluidized bed because of their very low density and granulometry. So, we didn't observe any enrichment of the bed in alkaline matters.

4 - Conclusions

The objective of the tests is now to determine more accurately the ideal working conditions for the fluidized bed reactor, and to establish, in those conditions, the mass and thermal balances and to determine more completely the composition of syngas with varying wood moisture and O_2 equivalence ratio.

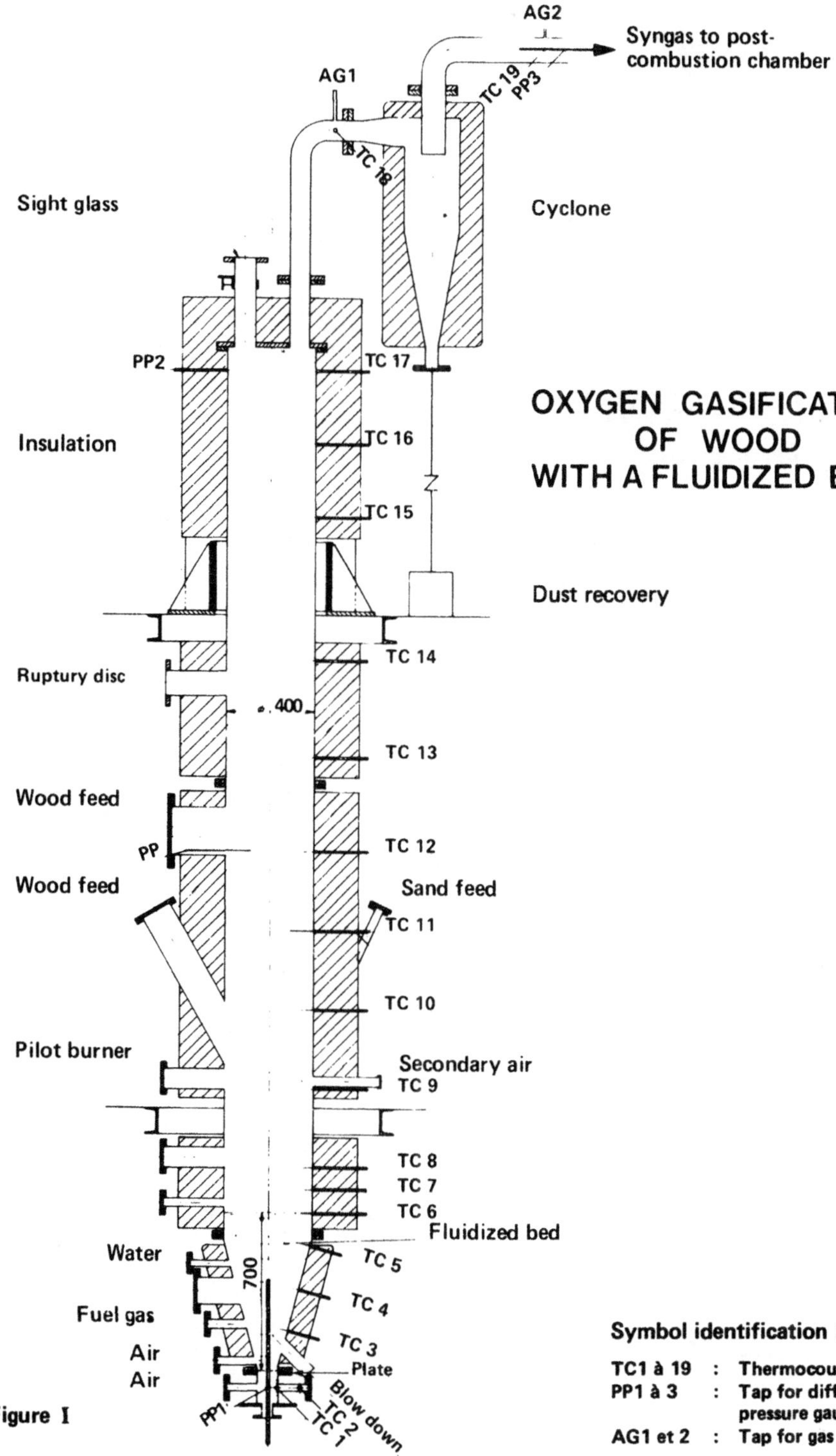

Symbol identification :

TC1 à 19	:	Thermocouples
PP1 à 3	:	Tap for differential pressure gauge
AG1 et 2	:	Tap for gas analysis

Figure I

gasifier overall view

PROGRESS IN THE DEVELOPMENT OF A TEST FACILITY FOR BIOMASS GASIFICATION STUDIES

Contract No.	: ES-E-R-032-UK(N)
Duration	: 24 months - August 1980-August 1982
Total Budget	: £256,000 C.E.C. contribution £91,400
Head of Project	: H.T. WILSON, Foster Wheeler Power Products Limited
Contractor	: Foster Wheeler Power Products Limited
Address	: Greater London House, Hampstead Road, London, NW1 7QN.

SUMMARY

The sanctioned work programme related to construction and operation of a 40 kg/hr. pressurised biomass gasification test facility, is well advanced. Commissioning of the completed plant is due to take place in August 1981.

A complementary theoretical study on gasification thermodynamics for the Carbon-Hydrogen-Oxygen system has been undertaken in order to target the gasifier operating conditions, towards maximising theoretical Methanol yield from the anticipated synthesis gas.

1.0 INTRODUCTION

In response to a 'call for tenders' from the C.E.C. (EEC No. C24317 - Solar Energy Programme, Sept. 1979) Foster Wheeler Power Products Limited (FWPP) submitted a proposal for a research programme related to the pressurised gasification of biomass. The proposed programme contained two main objectives :

(1) To examine the effects of process variations on the gasification of biomass to optimise the production of a synthesis gas as feed stock for Methanol production on large scale plants and,

(2) To compare the performance of a number of different gasifier reactors under similar process conditions and thus provide a basis on which the most effective reactor could be selected.

The sanctioned work programme is now well advanced, with significant developments in two key areas :

(i) Theoretical considerations and,

(ii) Design/Construction of test facility.

Due to procurement difficulties in particular related to the solids handling systems and design problems associated with the steam generator/superheater,the projected timetable for the programme of experimental work has altered. It is anticipated that commissioning of the facility will take place during August 1981.

A detailed theoretical assessment of gasification thermodynamics was initiated to support the proposed experimental programme. The results of this work have been used to target the initial operational conditions - thus acting to reduce the amount of experimental work necessary to optimise system performance. Application of the general theoretical models to existing coal gasification processes has resulted in clear validation due to the very high fit on plant output data.

2.0 THEORETICAL CONSIDERATIONS

FWPP have developed and modified existing C-H-O gas/solid equilibrium models to predict equilibrium gas compositions which would arise from the gasification of biomass.

Two areas have been extensively explored :

(a) Equilibrium with solid carbon present and

(b) Without solid carbon

With the assistance of a system of specially developed and very flexible computer programmes equilibrium condition of temperature, pressure and gasifying medium have been examined. Comprehensive 'maps' of the theoretically possible products have been produced.

The derived outputs have been checked against data available from the Texaco entrained bed gasifier, the FW-Stoic 2-stage gasifier, the U-Gas (IGT) gasifier and the Lurgi dry ash and slagging gasifiers. In all instances, predicted gas compositions have compared favourably with available experimentally derived results.

The results from the theoretical study (fig. 1) also appear to confirm that generally the equilibrium gas composition is invariant with the reaction pressure above approximately 10 bars. An 'optimum' system pressure is to be defined. The 'optimum' pressure will include such factors as equipment sizing, operational complexity, product gas compression and Methanol yield.

At present, FWPP is actively involved in relating the derived equilibrium gas compositions to Methanol yield, using a proprietary ICI Methanol loop programme. The results of this work will be presented in the next Contractors Meeting.

3.0 DESIGN/CONSTRUCTION

The detail design of the 40 kg/hr test plant posed a number of difficult problems which had to be resolved in order to progress the plant construction.

A major area of concern has been personnel safety. The combination of high pressure operation (up to 30 bar), high temperature (up to 1200°C), carbonaceous feed and toxic/flammable product gases (H_2, CO, CH_4 etc.) coupled with requirement for an oxygen supply facilty (oxy-propane burner) has resulted in a design which incorporates either by elimination or identification and control, a large number of safety features. The plant, which is at this time approaching completion, has been designed in accordance with ASME VIII-Division 1 procedures for an operating pressure of 30 bar.

The main problem areas were identified as :

(1) Solids Feeding/Removal
(2) Superheater Design.

In view of the toxic/flammable nature of the product gas, it is clear that it requires careful containment - this is a difficult constraint when faced with solids feeding into a pressurised vessel. The FWPP solution allows feed to be continuously batched into the reactor via a pressurised lock hopper and triple slide valve lock system. The lock interspaces are selectively pressurised/de-pressurised with inert gas. Feeding is undertaken on a semi-automatic basis with operator initiation of the cycle from a feed low signal taken from a nuclear level detector mounted on the reactor. Leakage gases are directed to flare. Ash/char products will be removed on a batch basis although provision is made for continuous batch discharge.

The feed system described is only suitable for operation pressures up to 10 bars, beyond this pressure, feed will be added on a batch basis.

The proposed operation of the plant allows for steam only gasification

3.0 (cont..)

- hence a requirement for superheated steam. Supply of superheated steam at the maximum temperature of 900°C proved to be a difficult proposition, because of complex heat transfer and mechanical considerations in the superheater tubes. The inherent problems were compounded by the need for variable steam rates demanded by the experimental programme. Rigorous analysis has resulted in a suitable design. This unit is presently under construction.

The facilities on the test plant include an on-line data logger to monitor and process instrumentation signals. The selected unit has the programme ability to alarm if set operation parameters are exceeded. A separate micro-computer dedicated to the data logger will be used to undertake continuous data analysis.

Design of alternative reactor configurations is in hand. The first reactor design is based on the FW-Cross Flow concept. The second configuration will incorporate a partial oxidation step, and the third will be based on the downdraught principle.

F.W.P.P.Ltd GASIFICATION WITH STEAM AND OXYGEN –
RD 695 EQUILIBRIUM. WITH NO SOLID CARBON

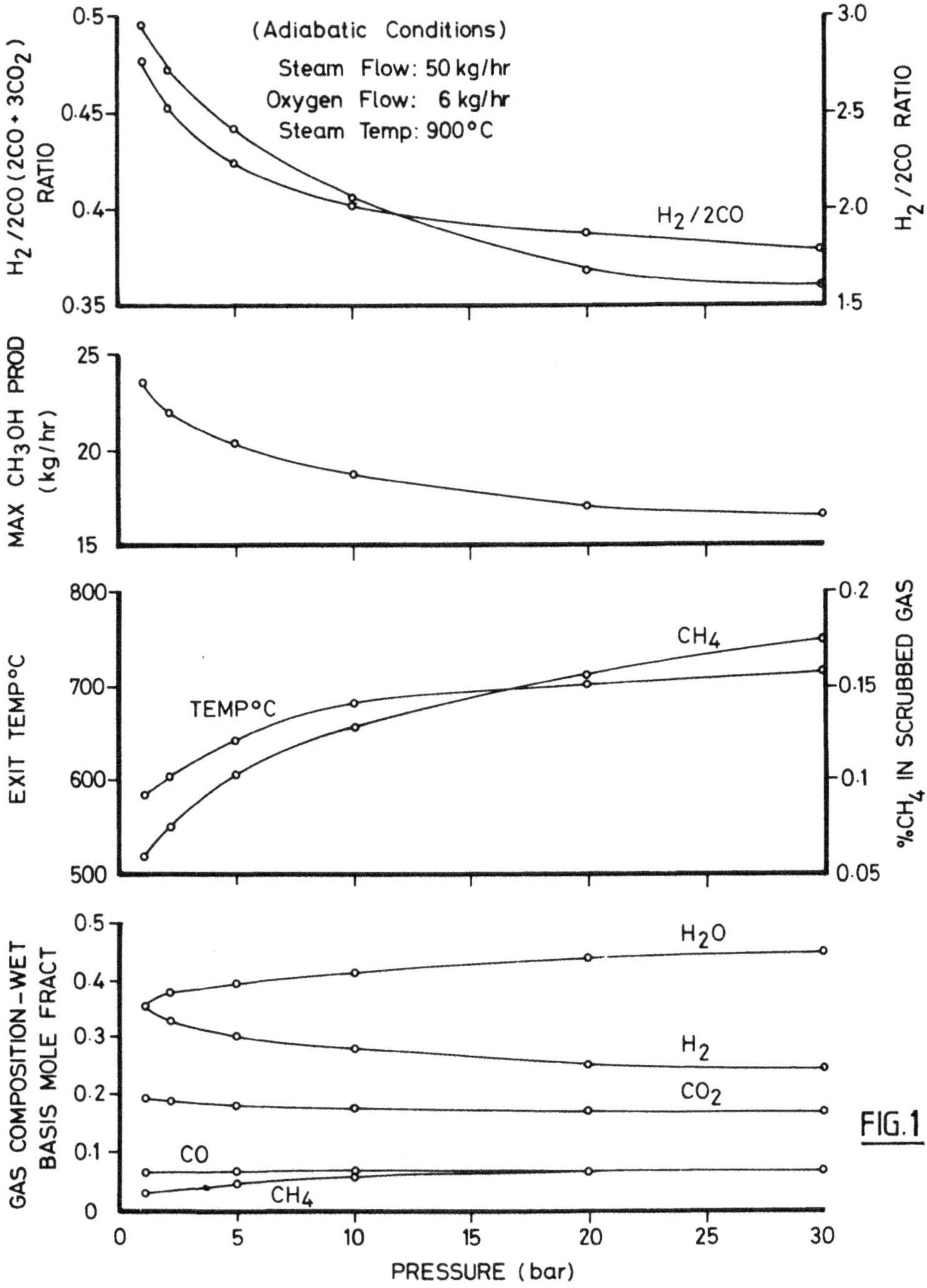

FIG.1

CATALYTIC LIQUEFACTION OF WOOD MATERIAL

Contract Number	: ESE-R-033-B
Duration	: 36 months 1 July 1980 - 30 June 1983
Total Budget	: 10.027.624 FB CEE Contribution: 100.000uc
Head of Project	: Prof. B. DELMON: Université Catholique de Louvain
Contractor	: Groupe de Physico-Chimie Minérale et de Catalyse
Address	: Place Croix du Sud 1 B-1348 Louvain-la-Neuve Belgium

Summary

The purpose of this work is to define the experimental conditions for the catalytic hydroliquefaction of woody material, with special emphasis on the choice of the catalytic systems and of the solvent which best achieve the total conversion of wood into valuable liquids either as fuel or chemicals (phenols). Orientation experiments reported hereafter were mainly devoted to a first evaluation of some important parameters, like temperature, hydrogen pressure, time of reaction, in the presence of Co-Mo hydrotreatment catalysts, and tetraline as hydrogen donor solvent. A second aspect of the present report concerns the different methods which are being used in order to gain informations on the classes of compounds which are present in the liquid phase collected after the reaction and possibly to identify the most abundant ones. Only partial results are presented.

1.1. Introduction

The objective of this programme is concerned with the catalytic degradation of the lignocellulosic constituents of the wood in reducing atmosphere, at temperatures ranging between 200-480°C, in view of obtaining fuels with satisfying characteristics and/or chemicals, in yields that are economically exploitable (essentially monophenols).

The informations actually available come from three main sources :

- attempts at producing oil by direct liquefaction of lignocellulose (or of cellulosic waste). These studies supply informations on the calorific value of the fuels, the evolution of the elemental compositions, material balance of the reaction, eventually some IR or NMR data of the crude.
- structural studies of lignin: they are most abundant and they are mainly concerned with the isolation of the monomers produced by controlled hydrogenolysis, in order to elucidate the nature of the elements which constitute lignin and their bondings. The main interest of those studies lie in the description of isolation and identification procedures. Indeed, the mild reaction conditions and the low yields are not comparable to those which prevail in semi-industrial processes. Fundamental research on the hydrogenation of oligosaccharides may be also mentioned.
- the research work on coal hydrogenation and solvent extraction. The experience accumulated in this domain may be most benefic, namely for the selection of appropriate catalytic systems and for the methods developed for the analysis of the complex organic mixtures obtained as reaction products.

1.2. Material, apparatus and analytical methods

1.2.1. Material : beech wood was used in all the experiments Its water content is 4% and was used as such.

1.2.2. Reaction apparatus and procedure. Two types of reactors are used.

. Preliminary tests are performed in a microreactor (∿ 10 ml). This equipment is mainly used for orientation runs where different catalysts, solvents and working conditions are screened. (Working conditions : 0-160 bar, 300°C).

. A 1 liter vessel (from Sotelem) allows to treat bigger wood masses (50-100 g), under pressures and temperatures up to 400 bar and 600°C. Continuous stirring of the slurry is maintained during the reaction. Sampling devices allow to follow the liquid and gas phase composition at different time intervals.

The experiments are generally performed on 200 g of slurry (wood + solvent). The vessel is purged with He before introducing hydrogen. The pressures are given by a pressure transducer. The temperature is measured by a thermocouple. The reactor is heated at a rate of 30°/min. The reaction temperature is reached after 20'. After reaction, the autoclave is cooled down to room temperature with compressed air.

1.2.3. Analytical procedures. The reaction products contain several organic compounds. The optimisation of the process with respect to the goal which is considered (hydrocarbons/ phenols) can only be effected after assessing a sufficient comprehension of the reaction in the experimental conditions, of the mechanisms and of the way the reaction is modified by the catalyst. This supposes, in a first step, the quantitative estimation of the different fractions which may provide informations on the nature of the chemical bonds that are preferentially broken, and the identification of the main reaction products within each fraction. In this respect, it is important to establish the following points : type of OH functions (phenolic or alcoholic), aromatic C/aliphatic C ratio, substitution degree, nature of the substituents, mean molecular mass etc... .

Bearing in mind those considerations, we have followed two analytical schemas.

a. Solvent extraction of the crude

The methodology is represented by the following schema (a.p. = aqueous phase; o.p. = organic phase)

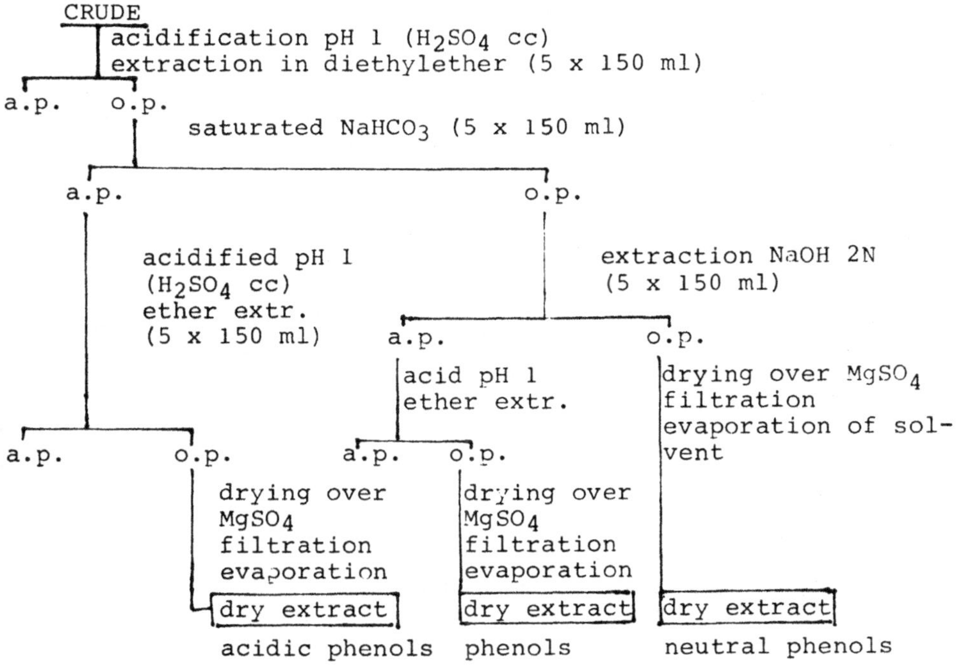

b. Fractional distillation of the crude

The crude is submitted to the following fractional distillation :

- distillation under atmospheric pressure between 50 and 95°C:"light fraction".
  between 95 and 100°C : water .
- distillation under reduced pressure (20-30 Torr) at 85°C (≃ 200°C at 1 bar) : fraction collected: solvent.

- distillation under 1.5-0.1 Torr, between 80-130°C (∿ 250-300°C under 1 bar): "heavy fraction" (monomeric phenols, aromatics, polycyclics).
- residue : "tar", soluble in acetone.

c. Identification methods.

The crude and the liquid fractions are analyzed using one or several of the following methods :

- gas-chromatography : (capillary column TP-Sil 5). The preliminary results obtained so far on the crude and on the different fractions are encouraging. In the near future a GC/MS will used to identify the different compounds.
- Infrared spectroscopy : is used for the identification of the main functional groups (OH, C=O, C=C aromatic, C=C aliphatic, CH, CH_2, CH_3 etc..).
- NMR spectroscopy (H and ^{13}C): provides informations on the presence of different OH groups (phenols, alcohols), mean C atom number per molecule, degree of aromaticity and substitution. The proeminent products may possibly be identified and quantified.
- CHO elemental analysis: preliminary analyses are in progress.

The systematic use of these methods should allow in the near future to draw some conclusions concerning the influence of the experimental parameters on the reaction. Other hydrogen donating solvents will be investigated and hydrogenation of simple molecules taken as models will also be carried out.

1.3. Results

1.3.1. Experimental parameters

- Catalysts. The experiments which have been carried out up to now aimed to test the active phases of hydrotreatment catalysts (Co-Mo sulfides) largely used in the petrochemical industry. Previous studies have shown that the hydrogenation and hydrogenolysis activities are maximum when the Co/Co+Mo = 0.05 and 0.3, respectively. They have been used in the first part of this work.
- Solvent : mainly tetraline (largely used for coal liquefaction).
- Temperature : 350-420°C.
- Hydrogen pressure : 0 - 85 bar
- Solvent/wood ratio : 3:1 (5:1)

1.3.2. Catalytic tests

- About 20 runs were carried out in the microreactor in order to precise the experimental conditions which best lead to the complete liquefaction of wood.
- The experimental conditions which were used with the 1 l pressure vessel are given in Table I.

In most cases, the liquefaction of wood was complete. The WHSV of the reaction is of the order of 40 h^{-1}, which is about 20 times as much as it is generally obtained in the hydrotreatment of petroleum crude.

TABLE I. Experiences carried out on 50 g of wood in 1 l pressure vessel. Experimental conditions: solvent : tetraline except run 1: H_2O, and run 2: cyclohexanol.

Run	T (°C)	Time (min)	P_{H_2} (bar)	$P_{H_2\,max}$ (bar)	Catalysts
1	340	60	60	165	Na_2CO_3 10%
2	350	60	60	174	-
3	395	15	60	169	CoMoS 5% Co 3.8 g
4	395	15	0	56	-
5	370	15	60	158	CoMoS 5% Co 4.0 g
6	420	15	60	204	CoMoS 5% Co 4.0 g
7	380	15	85	225	CoMoS 5% Co 4.0 g
8	380	15	60	172	-
9	400	15	30	115	CoMoS 5% Co 4.2 g
10*	420	15	60	158	CoMoS 5% Co 4.0 g
11	420	15	60	198	NiWS 30% Ni 4.5 g
12	420	30	60	191	CoMoS 5% Co 4.1 g
13	400	15	60	180	CoMoS30% Co 3.0 g
14	400	15	60	150	CoMoS30% Co 3.8 g
15	400	15	60	167	CoMoS30% Co 4.2 g
16	400	15	60	180	CoMoS30% Co 1.2 g
17**	400	15	60		CoMoS30% Co 4.0 g

* Solvent/wood : 5:1
** No wood

1.3.3. Analytical data of the liquid phase

a. Extraction procedure

The crudes of runs 1-4 were extracted according to the scheme given above in order to isolate the phenolic compounds.

Except for run 2 (performed with cyclohexanol) where 18% phenolic compounds were recovered (identification by ^{13}C NMR), the three other runs yielded $\sim$ 1% of phenols. No other identification was done.

b. Distillation procedure

This procedure allow to separate four classes of compounds: water, light and heavy oil, and tar (distillation residue). So far this procedure has been applied to 7 crudes and the data are given in Table II.

TABLE II. Fractional distillation: solvent recovery (%) and percentages of the different fractions expressed in g liq/100 g wood.

Run	% Solvent rec.	% Water	% Light	% Heavy	% Tar	Total liq. HC
6	88	2	8.5	2	20	30.5
7	63	∿ 0	0	1	5	6
8	45	2	2	13	6	21
9	94	10	6.8	∿ 0	23	29.8
10	100	28	18	3.6	32	53.6
11	86	18	4	5	33	42
17	100	0	0	0	0	0

Temporary observations can be made from these results.

1. Run 7: high P_{H_2} seems to favour gasification instead of liquefaction.
2. Run 8: in absence of catalyst, liquefaction occurs, which confirms the hydrogen donating properties of the solvent. Highest yield in heavy fraction.
3. Run 17: the solvent is not appreciably altered by the catalyst.
4. Run 6 and 10 : (similar experiment conditions): differ only by the solvent/wood ratio: higher yields are obtained with a 5:1 ratio.
5. Run 10: highest liquid % : 53.6, which is near the maximum so far reported in literature (∿ 57%).

N.B. Sulfidation of the catalyst is not controlled in the reactor.

c. Identification methods

Here again only fragmentary informations have been obtained at the present time. The set up of the analytical methods is in progress. In particular, the use of GCMS should enable us to identify soon the most important reaction products.

1.4. Conclusions

In the experimental conditions adopted in the first part of the study, complete liquefaction of wood is achieved.

Among the crudes so far examined, one crude contained over 50% liquid hydrocarbons, which is near the best conversion reported in literature from experiments carried out in similar conditions. But the tar fraction represents over 50% of this liquid phase.

On the basis of the results of these preliminary experiments, the main next step will be concerned with the choice of more active and selective catalytic systems.

A T T E N D A N C E   L I S T

CONTRACTORS' MEETING : Energy from Biomass - Project E

Copenhagen, 23 - 24 June 1981

NAME	INSTITUTION	ADDRESS	TEL.+ TELEX
BALLONI, W.	Centro Studi Microrganismi Autotrofi - Università Firenze	Piazzale delle Cascine 27 I - FIRENZE	055/36 05 06
BEENACKERS, A.A.C.M.	Twente University of Technology	P.O.Box 217 NL - 7500 AE ENSCHEDE	053/89 30 79
BELDMAN, G.	Food Science Dept. Agricult. University Wageningen	Biotechnion De Dreyen 12 NL - 6703 BC WAGENINGEN	08370/82 888
BERTHELSEN, L.	Jordbrugsteknisk Inst.	Agrovej 10 DK - 2630 TAASTRUP	02/9926 13
BRAUN, Ph.	INRA - Amélioration des Plantes	9, Place Viola F - 34060 MONTPELLIER	67/63 18 27
CALLAGHAN, T.V.	ITE (Natural Environm. Research Council) Cumbria	Merlewood GB - GRANGE-OVER-SANDS	044/84 2264
CANTARELLA, M.	University of Naples Istituto di Fisiologia	Via Mezzocannone 8 I - NAPOLI	081/61 18 00
CARRUTHERS, S.P.	Dept. of Agriculture + Horticultural Univ. of Reading	Earley Gate GB - READING	0734/66 15 18
CASADEVALL, E.	Ecole nationale sup. de Chimie Paris	11, r.P. et M.Curie F - 75231 PARIS CEDEX 05	1/336 25 25 Tx 3836
CHARTIER, Ph.	INRA - Dept. de Bio-Climatologie	route de St. Cyr F - 78000 VERSAILLES	950 75 22 Tx 695269f inraver
CHASSANY, J.P.	INRA - Economie rurale	ENSAM-CRAM Place Viala F - 34060 MONTPELLIER	67/63 19 32

NAME	INSTITUTION	ADDRESS	TEL.+ TELEX
COLLERAN, E.	University College Galway	EIR - GALWAY	
CONDON, L.	Forest and Wildlife Service	22, Upper Merrion St. EIR - DUBLIN 2	01/78 92 11
COOMBS, J.		Winchcombe Cottage Buclebury GB - READING	0635/62131
DAVIES, R.J.	Foster Wheeler Power Products	Greater London House Hampstead Rd. GB - LONDON NW1 7QN	01/388 12 12 Tx 263984
DUBOIS, P.	Laboratoires de Marcoussis - CRCGE	Route de Nozay F - 91 MARCOUSSIS	
ETHERIDGE, S.P.	University College Cardiff	Newport Rd. CARDIFF - S. Wales	0222/44 211
FARGET	EEC - DG VI/A2	200, rue de la Loi B - 1049 BRUSSELS (Ber. 513)	736 60 00 ext. 2590
FRANCESCO, A.	Istituto di Principi di Ingegneria Chimica Facoltà di Ingegneria	Piazzale Tecchio I - 80125 NAPOLI	081/61 18 00 081/61 09 66
GEHRMANN, J.	KFA Jülich - PLE	D - 517 JUELICH	02461/614852
GOMA, G.	Institut national des Sciences appliquées Toulouse	Av. Rangueil F - 31077 TOULOUSE	16(61)252113
GRAND	C.S.A.R.E.	Corso del Popolo 201 I - ROVIGO	0425/26478
GRASSI, G.	CEC - DG XII	200, rue de la Loi B - 1049 BRUSSELS	
GUDIN, C.	SFBP Lavera 13117	10, Quai Paul Doumer F - 92401 COURBEVOIE	42/81 08 78 ext. 723
HANSEN, G.K.	Dept. Plant Husbandry	Agrovej 10 DK - TAASTRUP	02/99 26 13
HANSEN, S.	OAC, Operations Analysis Centre A/S	Sofielundsvej 23 DK - 2600 GLOSTRUP	02/96 80 00 Tx 33 472 cabrop dk

NAME	INSTITUTION	ADDRESS	TEL.+ TELEX
HAVE, H.	Jordbrugsteknisk Inst.	Agrovej 10 DK - 2630 TAASTRUP	02/99 26 13
HEALY, J.	Bord Na Mona	Baggot Str. EIR - DUBLIN 2	688555 Tx 30206 mowa
HELARY, J.L.	Ateliers et Chantiers de Bretagne	CEDEX no. 2 F - 44040 NANTES CEDEX	40/47 31 32
HOLT, T.J.	University of Liverpool	Marine Biological Station PORT ERIN Isle of Man	44/624-892016
HURAND, A.	CEMAGREF	Parc de Tourvoie F - 92160 ANTONY	666 21 07
JASTER	Fritz Werner GmbH	D - GEISENHEIM	
JONES, J.M.(Ms)	University of Liverpool	Marine Biological Station PORT ERIN Isle of Man	44/624-892016
KING, J.	ETSU	AERE Harwell Didcot GB - OXON	0235/834621
LAFONTAINE, H.	H.T.S.	1995 Keystone Blvd USA - MIAMI Fl.33181	305/891 46 01
LARGEAU, C.	Ecole nationale sup. de Chimie de Paris CNRS	11, r. P.et M.Curie F - 75231 PARIS CEDEX 05	1/336 25 25 Tx 51 16
LEGROS, A.	Université de Louvain-la-Neuve	1, Pl. Croix du Sud Bte 9 B - 1348 LOUVAIN-LA NEUVE	010/41 81 81 ext. 3655
LETTINGA, G.	Agricultural University	De Dreyen 12 NL - WAGENINGEN	08370/83437
LICATA, R.	Phoebus	Via Ugo Ojetti 426 I - ROMA	
LINNEBORN, J.	Fritz Werner GmbH	D - GEISENHEIM	
MARCELLIN, M.	Creusot-Loire Division Energie	EN/PN/LF F - 71208 LE CREUSOT	85/55 80 80 ext. 75916 Tx 801280 f

NAME	INSTITUTION	ADDRESS	TEL.+ TELEX
MATERASSI, R.	Centro Studi Microrganismi Autotrofi - Università Firenze	Piazzale delle Cascine 27 I - FIRENZE	055/36 05 06
MITCHELL, C.P.	Forestry Dept. Aberdeen University	GB - ABERDEEN	
MOLLE, J.P.	CEMAGREF	Parc de Tourvoie F - 92160 ANTONY	666 21 09
MORLEY, J.G.	University of Nottingham	Wolfson Institute of Interfacial Technology	0602/56 101 ext. 3348
NAVEAU, H.	Université Catholique Louvain	1, Pl. Croix du Sud Bte 9 B - 1348 LOUVAIN-LA-NEUVE	010/41 81 81 ext. 3646
NEENAN, M.	An Foras Taluntais	Oakpark EIR - CARLOW	0503/31 425 Tx 33038
NEWALL, P.J.	University College Galway	EIR - GALWAY	
NICOLAY, D.	CEC - DG XIII/A2	B.P. 1907 L - LUXEMBOURG	43011 ext. 2946 Tx 3423/3446 comeur lu
NIELSEN, H.	Jordbrugsteknisk Inst.	Agrovej 10 DK - TAASTRUP	02/99 26 13
NIELSEN, H.K.	id.	id.	id.
NOACK, D.	Bundesforschungsanstalt für Forst- und Holzwirtschaft	Leuschnerstr. 91 D - 2050 HAMBURG 80	040/73 91 91
PALZ, W.	CEC - DG XII	200, rue de la Loi B - 1049 BRUSSELS	Tx 21877 comeu b
PEARCE, M.L.	Forestry Commission	TETBURY GB - GLOUCESTER	
PEDERSEN, T.T.	Jordbrugsteknisk Inst.	Rolighedsvej 23 DK - 1958 COPENHAGEN	01/351 788
PETER, F.	TOTAL	5, rue Michel-Ange F - 75016 PARIS	1/743 80 00 Tx 611992

NAME	INSTITUTION	ADDRESS	TEL.+ TELEX
PIRRWITZ,	CEC - DG XII	200, rue de la Loi B - 1049 BRUSSELS	Tx 21877 comeu b
PONCELET, C.	Univ.Catholique de Louvain - Gr. de Physico-Chimie minérale et de Catalyse	1, Pl. Croix du Sud B - 1348 LOUVAIN-LA NEUVE	010/41 81 81 ext. 3596
POULAIN, N.	Univ. Catholique de Louvain	B - 1348 LOUVAIN-LA NEUVE	010/41 81 81 ext. 3655
PUCCIONI-AGNOLETTI M.C.	Forestry Dept. Aberdeen University	GB - ABERDEEN	
RENARD, J.C.	Laboratoires de Marcoussis - CRCGE	Route de Nozay F - 91 MARCOUSSIS	449 11 72
RIJKENS, B.A.	Inst. for Storage and Processing of Agricultural Produce (IBVL)	Bornsesteeg 59 P.O.Box 18 NL - 6700 AA WAGENINGEN	08370/19043 Tx 45371
ROUSSE, M.	Creusot-Loire Entreprises	33, Quai Gallieni F - 92150 SURESNES	1/772 12 12
SCARAMUZZI, G.	Ente Nazionale Cellulosa e Carta	Via Casalotti 300 I - ROMA	06/696 02 41
SEVILA, F.	CEMAGREF	B.P. 4020 F - 30001 NIMES	66/84 84 79
SIRONVAL, C.	Université de Liège	Sart-Tilman B - 4000 LIEGE	041/56 18 26
STAFFORD, D.A.	Dept. of Microbiology University College Cardiff	Newport Rd. CARDIFF - S. Wales	0222 44211
TEISSIER DU CROS, E.	Institut national de la Recherche agronomique - Dept. Recherches forestières - Station d'amélioration des Arbres forestiers	Ardon F - 45160 OLIVET	38/63 15 35 Tx 760359f inraorl
VAN DEN AARSSEN, F.C.	Twente University of Technology	P.O.Box 217 NL - 7500 AE ENSCHEDE	053/89 30 28

NAME	INSTITUTION	ADDRESS	TEL.+ TELEX
VAN DER BURGT, N.J.	Shell Int. Petr. Maatschappij B.V.	Carel van Bijlandtlaan 30 NL - DEN HAAG	70/77 23 46 Tx 31005
VIANELLO, V.	C.S.A.R.E.	Corso del Popolo 201 I - ROVIGO	
VOETBERG, J.	Inst. for Storage and Processing of Agricultural Produce (IBVL)	Bornsesteeg 59 P.O.Box 18 NL - 6700 AA WAGENINGEN	08370/19043 Tx 45371
VORAGEN, A.G.Y.	Food Science Dept. Agricult. University Wageningen	Biotechnion De Dreyen 12 NL - 6703 BC WAGENINGEN	08370/82888
WAGENER, K.	Techn. Univ. Aachen	D - 51 AACHEN	02461/616711
WIEGANT, W.M.	Agricultural Univ. Dept. of Water Pollution Control	Salverdaplein 10 NL - 6701 DB WAGENINGEN	08370/83202
WILSON, H.T.	Foster Wheeler Power Products	Greater London House Hampstead Rd. GB - LONDON NW1 7QN	01/388 12 12 Tx 363984
WILTON, B.	University of Nottingham School of Agriculture	Sutton Bowington Nr. Loughborough GB - LEICESTERSHIRE	05097-2386

Batch number: 10761599

Printed by Printforce, the Netherlands